Understanding Fish Biology

Understanding Fish Biology

Brandon Hopkins

MURPHY & MOORE
www.murphy-moorepublishing.com

Murphy & Moore Publishing,
1 Rockefeller Plaza,
New York City, NY 10020, USA

Visit us on the World Wide Web at:
www.murphy-moorepublishing.com

ISBN: 978-1-63987-552-8 (Hardback)

Cataloging-in-Publication Data

Understanding fish biology / Brandon Hopkins.
p. cm.
Includes bibliographical references and index.
ISBN 978-1-63987-552-8
1. Fishes. 2. Fishes--Ecology. 3. Fishes--Habitat. 4. Aquatic biology.
5. Fisheries. I. Hopkins, Brandon.

QL615 .U53 2021
597--dc23

Table of Contents

	Preface	**VII**
Chapter 1	**Fish Biology: An Introduction**	**1**
	• Fish Development	23
	• Fish Habitats	30
	• Fish Migration	31
	• Fish Aggression	33
Chapter 2	**Fish Taxonomy**	**37**
	• Agnatha	38
	• Osteichthyes	43
	• Chondrichthyes	45
	• Sarcopterygii	55
Chapter 3	**Anatomical Physiology of Fishes**	**59**
	• Fish Physiology	59
	• Fish Jaw	75
	• Fish Gill	85
	• Fish Scales	89
	• Fish Fin	97
	• Fish Locomotion	113
	• Sensory System in Fish	126
	• Pain in Fish	131
	• Electroreception	146
	• Fish Intelligence	151
Chapter 4	**Reproduction in Fishes**	**161**
	• Fish Reproduction	161
	• Spawn	171
	• Ichthyoplankton	196
	• Pregnancy in Fish	199
Chapter 5	**Fish Diseases and Parasites**	**207**
	• Fish Disease	207
	• Bacterial Diseases of Fish	207

- Fungal Diseases of Fish 215
- Viral Diseases of Fish 217
- Fish Parasites and its Diseases 221

Permissions

Index

Preface

Fish are aquatic craniate animals which lack limbs with digits but possess gills. Some of their common types are bony fish, hagfish, lampreys and cartilaginous fish. These are cold blooded animals with the exception of active swimmers like sharks and tuna. Fish communicate with each other by means of acoustic communication. This involves production of stridulatory sounds and non-stridulatory sounds. Fish generate these sounds by moving parts of their skeletons or using a dedicated organ like the swimbladder. They exist in most of the aquatic environments ranging from high mountain streams to hadopelagic zones. Fish have streamlined bodies and have two sets of paired fins. Most of them are covered with scales and lay eggs. This book presents the complex subject of fish biology in the most comprehensible and easy to understand language. Its extensive content provides the readers with a thorough understanding of the subject. This textbook is appropriate for students seeking detailed information in this area as well as for experts.

Given below is the chapter wise description of the book:

Chapter 1- Fish biology is the branch of science which deals with the study of aquatic, craniate, gill-bearing animals that lack limbs with digits. Some of the topics studied under fish biology are fish development, fish habitats, fish migration and fish aggression. This is an introductory chapter which will introduce briefly all these significant aspects of fish biology.

Chapter 2- Fishes can be categorized into various classes such as Agnatha, Osteichthyes, Chondrichthyes and Sarcopterygii. Jawless fishes come under class Agnatha; Osteichthyes contains bony fishes, cartilaginous fishes fall under Chondrichthyes and Sarcopterygii contains fleshy finned fishes. This chapter has been carefully written to provide an easy understanding of these classes of fishes.

Chapter 3- Fish physiology deals with the study of the component of fishes as well as the way in which they function together. Some of the topics studied under anatomical physiology of fishes are fish jaw, fish gill, fish scale, fish fin, fish locomotion, sensory systems in fish, pain in fish, etc. The chapter closely examines these key aspects of anatomical physiology of fishes to provide an extensive understanding of the subject.

Chapter 4- Fishes are unisexual and reproduce sexually. Most fishes are oviparous and the fertilization is external. Many adult fishes of the same species come together in a group and release their gametes into the water at the same time. This process is known as spawning. The topics elaborated in this chapter will help in gaining a better perspective about the process reproduction in fishes.

Chapter 5- Fishes can suffer from numerous diseases and parasites. Some of the common diseases of fishes can be categorized as bacterial, fungal and viral diseases. This chapter closely examines these types of fish diseases as well as the parasites which can cause diseases in fishes.

At the end, I would like to thank all those who dedicated their time and efforts for the successful completion of this book. I also wish to convey my gratitude towards my friends and family who supported me at every step.

Brandon Hopkins

Chapter 1

Fish Biology: An Introduction

Fish biology is the branch of science which deals with the study of aquatic, craniate, gill-bearing animals that lack limbs with digits. Some of the topics studied under fish biology are fish development, fish habitats, fish migration and fish aggression. This is an introductory chapter which will introduce briefly all these significant aspects of fish biology.

The biology of fishes is the biology of active living organisms in water. The basics of fish biology, which entails are Anatomy, Physiology, Embryology and Endocrinology of bony and cartilaginous fish. Fishes are cold blooded or poikilothermic animals i.e. their body temperature varying passively in accordance with the ambient temperature (surrounding water temperature). Although, fishes as a group can tolerate wide range of temperature just below 0 °C to 45 °C, individual species generally have a preferred or optimum as well as a more restricted temperature range. Fisheries science explores the interaction among fish populations, the environment, and humans. Fisheries researchers confront issues ranging from managing harvest of fish populations to developing novel approaches for culturing finfish and shellfish.

Fish

Fish is any of approximately 34,000 species of vertebrate animals (phylum Chordata) found in the fresh and salt waters of the world. Living species range from the primitive jawless lampreys and hagfishes through the cartilaginous sharks, skates, and rays to the abundant and diverse bony fishes. Most fish species are cold-blooded; however, one species, the opah (Lampris guttatus), is warm-blooded.

Pumpkinseed sunfish (*Lepomis gibbosus*).

The term *fish* is applied to a variety of vertebrates of several evolutionary lines. It describes a life-form rather than a taxonomic group. As members of the phylum Chordata, fish share certain features with other vertebrates. These features are gill slits at some point in the life cycle, a notochord, or skeletal supporting rod, a dorsal hollow nerve cord, and a tail. Living fishes represent some five classes, which are as distinct from one another as are the four classes of familiar air-breathing animals—amphibians, reptiles, birds, and mammals. For example, the jawless fishes (Agnatha) have gills in pouches and lack limb girdles. Extant agnathans are the lampreys and the hagfishes. As the name implies, the skeletons of fishes of the class Chondrichthyes (from *chondr*, "cartilage," and *ichthyes*, "fish") are made entirely of cartilage. Modern fish of this class lack a swim bladder, and their scales and teeth are made up of the same placoid material. Sharks, skates, and rays are examples of cartilaginous fishes. The bony fishes are by far the largest class. Examples range from the tiny sea horse to the 450-kg (1,000-pound) blue marlin, from the flattened soles and flounders to the boxy puffers and ocean sunfishes. Unlike the scales of the cartilaginous fishes, those of bony fishes, when present, grow throughout life and are made up of thin overlapping plates of bone. Bony fishes also have an operculum that covers the gill slits.

Lamprey Lamprey (*Lampetra*) on rainbow trout.

The study of fishes, the science of ichthyology, is of broad importance. Fishes are of interest to humans for many reasons, the most important being their relationship with and dependence on the environment. A more obvious reason for interest in fishes is their role as a moderate but important part of the world's food supply. This resource, once thought unlimited, is now realized to be finite and in delicate balance with the biological, chemical, and physical factors of the aquatic environment. Overfishing, pollution, and alteration of the environment are the chief enemies of proper fisheries management, both in fresh waters and in the ocean. Another practical reason for studying fishes is their use in disease control. As predators on mosquito larvae, they help curb malaria and other mosquito-borne diseases.

Landing a fish catch in the harbour of Esbjerg, Den.

Fishes are valuable laboratory animals in many aspects of medical and biological research. For example, the readiness of many fishes to acclimate to captivity has allowed biologists to study behaviour, physiology, and even ecology under relatively natural conditions. Fishes have been especially important in the study of animal behaviour, where research on fishes has provided a broad base for the understanding of the more flexible behaviour of the higher vertebrates. The zebra fish is used as a model in studies of gene expression.

There are aesthetic and recreational reasons for an interest in fishes. Millions of people keep live fishes in home aquariums for the simple pleasure of observing the beauty and behaviour of animals otherwise unfamiliar to them. Aquarium fishes provide a personal challenge to many aquarists, allowing them to test their ability to keep a small section of the natural environment in their homes. Sportfishing is another way of enjoying the natural environment, also indulged in by millions of people every year. Interest in aquarium fishes and sportfishing supports multimillion-dollar industries throughout the world.

Harlequin fish (*Rasbora heteromorpha*).

General Features

Structural Diversity

Fishes have been in existence for more than 450 million years, during which time they have evolved repeatedly to fit into almost every conceivable type of aquatic habitat. In a sense, land vertebrates are simply highly modified fishes: when fishes colonized the land habitat, they became tetrapod (four-legged) land vertebrates. The popular conception of a fish as a slippery, streamlined aquatic animal that possesses fins and breathes by gills applies to many fishes, but far more fishes deviate from that conception than conform to it. For example, the body is elongate in many forms and greatly shortened in others; the body is flattened in some (principally in bottom-dwelling fishes) and laterally compressed in many others; the fins may be elaborately extended, forming intricate shapes, or they may be reduced or even lost; and the positions of the mouth, eyes, nostrils, and gill openings vary widely. Air breathers have appeared in several evolutionary lines.

Trumpet fish (*Aulostomus maculatus*).

Many fishes are cryptically coloured and shaped, closely matching their respective environments; others are among the most brilliantly coloured of all organisms, with a wide range of hues, often of striking intensity, on a single individual. The brilliance of pigments may be enhanced by the surface structure of the fish, so that it almost seems to glow. A number of unrelated fishes have actual light-producing organs. Many fishes are able to alter their coloration—some for the purpose of camouflage, others for the enhancement of behavioral signals.

Oyster toadfish (*Opsanus tau*).

Fishes range in adult length from less than 10 mm (0.4 inch) to more than 20 metres (60 feet) and in weight from about 1.5 grams (less than 0.06 ounce) to many thousands of kilograms. Some live in shallow thermal springs at temperatures slightly above 42 °C (100 °F), others in cold Arctic seas a few degrees below 0 °C (32 °F) or in cold deep waters more than 4,000 metres (13,100 feet) beneath the ocean surface. The structural and, especially, the physiological adaptations for life at such extremes are relatively poorly known and provide the scientifically curious with great incentive for study.

Evolution and Paleontology

Almost all natural bodies of water bear fish life, the exceptions being very hot thermal ponds and extremely salt-alkaline lakes, such as the Dead Sea in Asia and the Great Salt Lake in North America. The present distribution of fishes is a result of the geological history and development of Earth as well as the ability of fishes to undergo evolutionary change and to adapt to the available habitats. Fishes may be seen to be distributed according to habitat and according to geographical area. Major habitat differences are marine and freshwater. For the most part, the fishes in a marine habitat differ from those in a freshwater habitat, even in adjacent areas, but some, such as the salmon, migrate from one to the other. The freshwater habitats may be seen to be of many kinds. Fishes found in mountain torrents, Arctic lakes, tropical lakes, temperate streams, and tropical rivers will all differ from each other, both in obvious gross structure and in physiological attributes. Even in closely adjacent habitats where, for example, a tropical mountain torrent enters a lowland stream, the fish fauna will differ. The marine habitats can be divided into deep ocean floors (benthic), mid-water oceanic (bathypelagic), surface oceanic (pelagic), rocky coast, sandy coast, muddy shores, bays, estuaries, and others. Also, for example, rocky coastal shores in tropical and temperate regions will have different fish faunas, even when such habitats occur along the same coastline.

Although much is known about the present geographical distribution of fishes, far less is known about how that distribution came about. Many parts of the fish fauna of the fresh waters of

North America and Eurasia are related and undoubtedly have a common origin. The faunas of Africa and South America are related, extremely old, and probably an expression of the drifting apart of the two continents. The fauna of southern Asia is related to that of Central Asia, and some of it appears to have entered Africa. The extremely large shore-fish faunas of the Indian and tropical Pacific oceans comprise a related complex, but the tropical shore fauna of the Atlantic, although containing Indo-Pacific components, is relatively limited and probably younger. The Arctic and Antarctic marine faunas are quite different from each other. The shore fauna of the North Pacific is quite distinct, and that of the North Atlantic more limited and probably younger. Pelagic oceanic fishes, especially those in deep waters, are similar the world over, showing little geographical isolation in terms of family groups. The deep oceanic habitat is very much the same throughout the world, but species differences do exist, showing geographical areas determined by oceanic currents and water masses.

All aspects of the life of a fish are closely correlated with adaptation to the total environment, physical, chemical, and biological. In studies, all the interdependent aspects of fish, such as behaviour, locomotion, reproduction, and physical and physiological characteristics, must be taken into account.

Correlated with their adaptation to an extremely wide variety of habitats is the extremely wide variety of life cycles that fishes display. The great majority hatch from relatively small eggs a few days to several weeks or more after the eggs are scattered in the water. Newly hatched young are still partially undeveloped and are called larvae until body structures such as fins, skeleton, and some organs are fully formed. Larval life is often very short, usually less than a few weeks, but it can be very long, some lampreys continuing as larvae for at least five years. Young and larval fishes, before reaching sexual maturity, must grow considerably, and their small size and other factors often dictate that they live in a habitat different than that of the adults. For example, most tropical marine shore fishes have pelagic larvae. Larval food also is different, and larval fishes often live in shallow waters, where they may be less exposed to predators.

After a fish reaches adult size, the length of its life is subject to many factors, such as innate rates of aging, predation pressure, and the nature of the local climate. The longevity of a species in the protected environment of an aquarium may have nothing to do with how long members of that species live in the wild. Many small fishes live only one to three years at the most. In some species, however, individuals may live as long as 10 or 20 or even 100 years.

Behaviour

Fish behaviour is a complicated and varied subject. As in almost all animals with a central nervous system, the nature of a response of an individual fish to stimuli from its environment depends upon the inherited characteristics of its nervous system, on what it has learned from past experience, and on the nature of the stimuli. Compared with the variety of human responses, however, that of a fish is stereotyped, not subject to much modification by "thought" or learning, and investigators must guard against anthropomorphic interpretations of fish behaviour.

Fishes perceive the world around them by the usual senses of sight, smell, hearing, touch, and taste and by special lateral line water-current detectors. In the few fishes that generate electric fields, a process that might best is called electrolocation aids in perception. One or another of these senses often is emphasized at the expense of others, depending upon the fish's other adaptations. In fishes with large eyes, the sense of smell may be reduced; others, with small eyes, hunt and feed primarily by smell (such as some eels).

Specialized behaviour is primarily concerned with the three most important activities in the fish's life: feeding, reproduction, and escape from enemies. Schooling behaviour of sardines on the high seas, for instance, is largely a protective device to avoid enemies, but it is also associated with and modified by their breeding and feeding requirements. Predatory fishes are often solitary, lying in wait to dart suddenly after their prey, a kind of locomotion impossible for beaked parrot fishes, which feed on coral, swimming in small groups from one coral head to the next. In addition, some predatory fishes that inhabit pelagic environments, such as tunas, often school.

Sleep in fishes, all of which lack true eyelids, consists of a seemingly listless state in which the fish maintains its balance but moves slowly. If attacked or disturbed, most can dart away. A few kinds of fishes lie on the bottom to sleep. Most catfishes, some loaches, and some eels and electric fishes are strictly nocturnal, being active and hunting for food during the night and retiring during the day to holes, thick vegetation, or other protective parts of the environment.

Communication between members of a species or between members of two or more species often is extremely important, especially in breeding behaviour (see below Reproduction). The mode of communication may be visual, as between the small so-called cleaner fish and a large fish of a very different species. The larger fish often allows the cleaner to enter its mouth to remove gill parasites. The cleaner is recognized by its distinctive colour and actions and therefore is not eaten, even if the larger fish is normally a predator. Communication is often chemical, signals being sent by specific chemicals called pheromones.

Locomotion

Many fishes have a streamlined body and swim freely in open water. Fish locomotion is closely correlated with habitat and ecological niche (the general position of the animal to its environment).

Tropical two-wing flying fish: Flying fishes, such as the tropical two-wing flying fish (*Exocoetus volitans*), do not fly in the sense of flapping their wing-sized fins. Their fins do allow them to glide in the air, however, after building up enough speed from swimming to break the surface of the water.

Many fishes in both marine and fresh waters swim at the surface and have mouths adapted to feed best (and sometimes only) at the surface. Often such fishes are long and slender, able to dart at surface insects or at other surface fishes and in turn to dart away from predators; needlefishes, halfbeaks, and topminnows (such as killifish and mosquito fish) are good examples. Oceanic flying fishes escape their predators by gathering speed above the water surface, with the lower lobe of the tail providing thrust in the water. They then glide hundreds of yards on enlarged, winglike pectoral and pelvic fins. South American freshwater flying fishes escape their enemies by jumping and propelling their strongly keeled bodies out of the water.

Mosquitofish (*Gambusia affinis*). The mosquitofish (*Gambusia affinis*), a type of topminnow native to the fresh waters of the southeastern United States, bears its young alive.

So-called mid-water swimmers, the most common type of fish, are of many kinds and live in many habitats. The powerful fusiform tunas and the trouts, for example, are adapted for strong, fast swimming, the tunas to capture prey speedily in the open ocean and the trouts to cope with the swift currents of streams and rivers. The trout body form is well adapted to many habitats. Fishes that live in relatively quiet waters such as bays or lake shores or slow rivers usually are not strong, fast swimmers but are capable of short, quick bursts of speed to escape a predator. Many of these fishes have their sides flattened, examples being the sunfish and the freshwater angelfish of aquarists. Fish associated with the bottom or substrate usually are slow swimmers. Open-water plankton-feeding fishes almost always remain fusiform and are capable of rapid, strong movement (for example, sardines and herrings of the open ocean and also many small minnows of streams and lakes).

Bluefin tuna (*Thunnus thynnus*).

Bottom-living fishes are of many kinds and have undergone many types of modification of their body shape and swimming habits. Rays, which evolved from strong-swimming mid-water sharks, usually stay close to the bottom and move by undulating their large pectoral fins. Flounders live in a similar habitat and move over the bottom by undulating the entire body. Many bottom fishes dart from place to place, resting on the bottom between movements, a motion common in gobies. One goby relative, the mudskipper, has taken to living at the edge of pools along the shore of muddy mangrove swamps. It escapes its enemies by flipping rapidly over the mud, out of the water. Some catfishes, synbranchid eels, the so-called climbing perch, and a few other fishes venture out over damp ground to find more promising waters than those that they left. They move by wriggling their bodies, sometimes using strong pectoral fins; most have accessory air-breathing organs. Many bottom-dwelling fishes live in mud holes or rocky crevices. Marine eels and gobies commonly are found in such habitats and for the most part venture far beyond their cave like homes. Some bottom dwellers, such as the clingfishes (Gobiesocidae), have developed powerful adhesive disks that enable them to remain in place on the substrate in areas such as rocky coasts, where the action of the waves is great.

Southern stingray (*Dasyatis americana* or *Hypanus americanus*): Southern stingrays (*Dasyatis americana* or *Hypanus americanus*) are found in the western Atlantic Ocean.

Reproduction

The methods of reproduction in fishes are varied, but most fishes lay a large number of small eggs, fertilized and scattered outside of the body. The eggs of pelagic fishes usually remain suspended in the open water. Many shore and freshwater fishes lay eggs on the bottom or among plants. Some have adhesive eggs. The mortality of the young and especially of the eggs is very high, and often only a few individuals grow to maturity out of hundreds, thousands, and in some cases millions of eggs laid.

Males produce sperm, usually as a milky white substance called milt, in two (sometimes one) testes within the body cavity. In bony fishes a sperm duct leads from each testis to a urogenital opening behind the vent or anus. In sharks and rays and in cyclostomes the duct leads to a cloaca. Sometimes the pelvic fins are modified to help transmit the milt to the eggs at the female's vent or on the substrate where the female has placed them. Sometimes accessory organs are used to fertilize females internally—for example, the claspers of many sharks and rays.

In the females the eggs are formed in two ovaries (sometimes only one) and pass through the ovaries to the urogenital opening and to the outside. In some fishes the eggs are fertilized internally but are shed before development takes place. Members of about a dozen families each of bony fishes (teleosts) and sharks bear live young. Many skates and rays also bear live young. In some bony fishes the eggs simply develop within the female, the young emerging when the eggs hatch (ovoviviparous). Others develop within the ovary and are nourished by ovarian tissues after hatching (viviparous). There are also other methods utilized by fishes to nourish young within the female. In all live-bearers the young are born at a relatively large size and are few in number. In one family of primarily marine fishes, the surfperches from the Pacific coast of North America, Japan, and Korea, the males of at least one species are born sexually mature, although they are not fully grown. Some fishes are hermaphroditic—an individual producing both sperm and eggs, usually at different stages of its life. Self-fertilization, however, is probably rare.

Successful reproduction and, in many cases, defense of the eggs and the young are assured by rather stereotypical but often elaborate courtship and parental behaviour, either by the male or the female or both. Some fishes prepare nests by hollowing out depressions in the sand bottom (cichlids, for example), build nests with plant materials and sticky threads excreted by the kidneys (sticklebacks), or blow a cluster of mucus-covered bubbles at the water surface (gouramis). The eggs are laid in these structures. Some varieties of cichlids and catfishes incubate eggs in their mouths.

Some fishes, such as salmon, undergo long migrations from the ocean and up large rivers to spawn in the gravel beds where they themselves hatched (anadromous fishes). Some, such as the freshwater eels (family Anguillidae), live and grow to maturity in fresh water and migrate to the sea to spawn (catadromous fishes). Other fishes undertake shorter migrations from lakes into streams, within the ocean, or enter spawning habitats that they do not ordinarily occupy in other ways.

Form and Function

Body Plan

The basic structure and function of the fish body are similar to those of all other vertebrates. The usual four types of tissues are present: surface or epithelial, connective (bone, cartilage, and fibrous tissues, as well as their derivative, blood), nerve, and muscle tissues. In addition, the fish's organs and organ systems parallel those of other vertebrates.

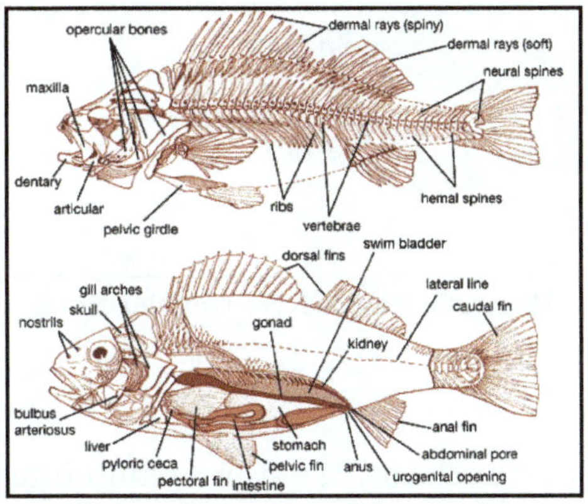

Internal structure of fishes. (Top) Skeleton of a perch (order Perciformes). (Bottom) Dissection of a perch.

The typical fish body is streamlined and spindle-shaped, with an anterior head, a gill apparatus, and a heart, the latter lying in the midline just below the gill chamber. The body cavity, containing the vital organs, is situated behind the head in the lower anterior part of the body. The anus usually marks the posterior termination of the body cavity and most often occurs just in front of the base of the anal fin. The spinal cord and vertebral column continue from the posterior part of the head to the base of the tail fin, passing dorsal to the body cavity and through the caudal (tail) region behind the body cavity. Most of the body is of muscular tissue, a high proportion of which is necessitated by swimming. In the course of evolution this basic body plan has been modified repeatedly into the many varieties of fish shapes that exist today.

The skeleton forms an integral part of the fish's locomotion system, as well as serving to protect vital parts. The internal skeleton consists of the skull bones (except for the roofing bones of the head, which are really part of the external skeleton), the vertebral column, and the fin supports (fin rays). The fin supports are derived from the external skeleton but will be treated here because of their close functional relationship to the internal skeleton. The internal skeleton of cyclostomes, sharks, and rays is of cartilage; that of many fossil groups and some primitive living fishes is mostly of cartilage but may include some bone. In place of the vertebral column, the earliest vertebrates had a fully developed notochord, a flexible stiff rod of viscous cells surrounded by a strong fibrous sheath. During the evolution of modern fishes the rod was replaced in part by cartilage and then by ossified cartilage. Sharks and rays retain a cartilaginous vertebral column; bony fishes have spool-shaped vertebrae that in the more primitive living forms only partially replace the notochord. The skull, including the gill arches and jaws of bony fishes, is fully, or at least partially, ossified. That of sharks and rays remains cartilaginous, at times partially replaced by calcium deposits but never by true bone.

The supportive elements of the fins (basal or radial bones or both) have changed greatly during fish evolution. Some of these changes are described in the section below (Evolution and paleontology). Most fishes possess a single dorsal fin on the midline of the back. Many have two and a few have three dorsal fins. The other fins are the single tail and anal fins and paired pelvic and pectoral fins. A small fin, the adipose fin, with hairlike fin rays, occurs in many of the relatively primitive teleosts (such as trout) on the back near the base of the caudal fin.

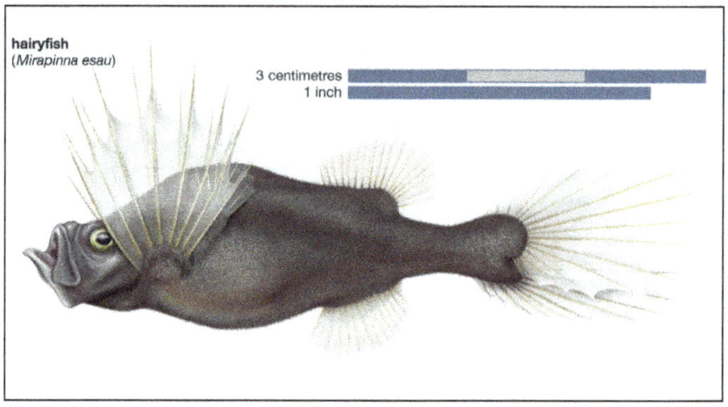

Hairyfish. Hairyfish (*Mirapinna esau*).

The Skin

The skin of a fish must serve many functions. It aids in maintaining the osmotic balance, provides physical protection for the body, is the site of coloration, contains sensory receptors, and, in some fishes, functions in respiration. Mucous glands, which aid in maintaining the water balance and offer protection from bacteria, are extremely numerous in fish skin, especially in cyclostomes and teleosts. Since mucous glands are present in the modern lampreys, it is reasonable to assume that they were present in primitive fishes, such as the ancient Silurian and Devonian agnathans. Protection from abrasion and predation is another function of the fish skin, and dermal (skin) bone arose early in fish evolution in response to this need. It is thought that bone first evolved in skin and only later invaded the cartilaginous areas of the fish's body, to provide additional support and protection. There is some argument as to which came first, cartilage or bone, and fossil evidence does not settle the question. In any event, dermal bone has played an important part in fish evolution and has different characteristics in different groups of fishes. Several groups are characterized at least in part by the kind of bony scales they possess.

Scales have played an important part in the evolution of fishes. Primitive fishes usually had thick bony plates or thick scales in several layers of bone, enamel, and related substances. Modern teleost fishes have scales of bone, which, while still protective, allow much more freedom of motion in the body. A few modern teleosts (some catfishes, sticklebacks, and others) have secondarily acquired bony plates in the skin. Modern and early sharks possessed placoid scales, a relatively primitive type of scale with a toothlike structure, consisting of an outside layer of enamel-like substance (vitrodentine), an inner layer of dentine, and a pulp cavity containing nerves and blood vessels. Primitive bony fishes had thick scales of either the ganoid or the cosmoid type. Cosmoid scales have a hard, enamel-like outer layer, an inner layer of cosmine (a form of dentine), and then a layer of vascular bone (isopedine). In ganoid scales the hard outer layer is different chemically and is called ganoin. Under this is a cosminelike layer and then a vascular bony layer. The thin, translucent bony scales of modern fishes, called cycloid and ctenoid (the latter distinguished by serrations at the edges), lack enameloid and dentine layers.

Skin has several other functions in fishes. It is well supplied with nerve endings and presumably receives tactile, thermal, and pain stimuli. Skin is also well supplied with blood vessels. Some fishes breathe in part through the skin, by the exchange of oxygen and carbon dioxide between the surrounding water and numerous small blood vessels near the skin surface.

Skin serves as protection through the control of coloration. Fishes exhibit an almost limitless range of colours. The colours often blend closely with the surroundings, effectively hiding the animal. Many fishes use bright colours for territorial advertisement or as recognition marks for other members of their own species, or sometimes for members of other species. Many fishes can change their colour to a greater or lesser degree, by movement of pigment within the pigment cells (chromatophores). Black pigment cells (melanophores), of almost universal occurrence in fishes, are often juxtaposed with other pigment cells. When placed beneath iridocytes or leucophores (bearing the silvery or white pigment guanine), melanophores produce structural colours of blue and green. These colours are often extremely intense, because they are formed by refraction of light through the needlelike crystals of guanine. The blue and green refracted colours are often relatively pure, lacking the red and yellow rays, which have been absorbed by the black pigment (melanin) of the melanophores. Yellow, orange, and red colours are produced by erythrophores, cells containing the appropriate carotenoid pigments. Other colours are produced by combinations of melanophores, erythrophores, and iridocytes.

Rosy rockfish (*Sebastes rosaceus*).

The Muscle System

The major portion of the body of most fishes consists of muscles. Most of the mass is trunk musculature, the fin muscles usually being relatively small. The caudal fin is usually the most powerful fin, being moved by the trunk musculature. The body musculature is usually arranged in rows of chevron-shaped segments on each side. Contractions of these segments, each attached to adjacent vertebrae and vertebral processes, bends the body on the vertebral joint, producing successive undulations of the body, passing from the head to the tail, and producing driving strokes of the tail. It is the latter that provides the strong forward movement for most fishes.

The Digestive System

The digestive system, in a functional sense, starts at the mouth, with the teeth used to capture prey or collect plant foods. Mouth shape and tooth structure vary greatly in fishes, depending on the kind of food normally eaten. Most fishes are predacious, feeding on small invertebrates or other fishes and have simple conical teeth on the jaws, on at least some of the bones of the roof of the mouth, and on special gill arch structures just in front of the esophagus. The latter are throat teeth. Most predacious fishes swallow their prey whole, and the teeth are used for grasping and holding prey, for orienting prey to be swallowed (head first) and for working the prey toward the esophagus. There are a variety of tooth types in fishes. Some fishes, such as sharks and piranhas, have cutting teeth for biting chunks out of their victims. A shark's tooth, although superficially like that of a piranha, appears in many respects to be a modified scale, while that of the piranha is like that of other bony fishes, consisting of dentine and enamel. Parrot fishes have beaklike mouths with short incisor-like teeth for breaking off coral and have heavy pavementlike throat teeth for crushing the coral. Some catfishes have small brushlike teeth, arranged in rows on the jaws, for scraping plant and animal growth from rocks. Many fishes (such as the Cyprinidae or minnows) have no jaw teeth at all but have very strong throat teeth.

Some fishes gather planktonic food by straining it from their gill cavities with numerous elongate stiff rods (gill rakers) anchored by one end to the gill bars. The food collected on these rods is passed to the throat, where it is swallowed. Most fishes have only short gill rakers that help keep food particles from escaping out the mouth cavity into the gill chamber.

Once reaching the throat, food enters a short, often greatly distensible esophagus, a simple tube with a muscular wall leading into a stomach. The stomach varies greatly in fishes, depending upon the diet. In most predacious fishes it is a simple straight or curved tube or pouch with a muscular wall and a glandular lining. Food is largely digested there and leaves the stomach in liquid form.

Between the stomach and the intestine, ducts enter the digestive tube from the liver and pancreas. The liver is a large, clearly defined organ. The pancreas may be embedded in it, diffused through it, or broken into small parts spread along some of the intestine. The junction between the stomach and the intestine is marked by a muscular valve. Pyloric ceca (blind sacs) occur in some fishes at this junction and have a digestive or absorptive function or both.

The intestine itself is quite variable in length, depending upon the fish's diet. It is short in predacious forms, sometimes no longer than the body cavity, but long in herbivorous forms, being coiled and several times longer than the entire length of the fish in some species of South American catfishes. The intestine is primarily an organ for absorbing nutrients into the bloodstream.

The larger its internal surface, the greater it's absorptive efficiency and a spiral valve is one method of increasing its absorption surface.

Sharks, rays, chimaeras, lungfishes, surviving chondrosteans, holosteans, and even a few of the more primitive teleosts have a spiral valve or at least traces of it in the intestine. Most modern teleosts have increased the area of the intestinal walls by having numerous folds and villi (finger like projections) somewhat like those in humans. Undigested substances are passed to the exterior through the anus in most teleost fishes. In lungfishes, sharks, and rays, it is first passed through the cloaca, a common cavity receiving the intestinal opening and the ducts from the urogenital system.

The Respiratory System

Oxygen and carbon dioxide dissolve in water, and most fishes exchange dissolved oxygen and carbon dioxide in water by means of the gills. The gills lie behind and to the side of the mouth cavity and consist of fleshy filaments supported by the gill arches and filled with blood vessels, which give gills a bright red colour. Water taken in continuously through the mouth passes backward between the gill bars and over the gill filaments, where the exchange of gases takes place. The gills are protected by a gill cover in teleosts and many other fishes but by flaps of skin in sharks, rays, and some of the older fossil fish groups. The blood capillaries in the gill filaments are close to the gill surface to take up oxygen from the water and to give up excess carbon dioxide to the water.

Most modern fishes have a hydrostatic (ballast) organ, called the swim bladder that lies in the body cavity just below the kidney and above the stomach and intestine. It originated as a diverticulum of the digestive canal. In advanced teleosts, especially the acanthopterygians, the bladder has lost its connection with the digestive tract, a condition called physoclistic. The connection has been retained (physostomous) by many relatively primitive teleosts. In several unrelated lines of fishes, the bladder has become specialized as a lung or, at least, as a highly vascularized accessory breathing organ. Some fishes with such accessory organs are obligate air breathers and will drown if denied access to the surface, even in well-oxygenated water. Fishes with a hydrostatic form of swim bladder can control their depth by regulating the amount of gas in the bladder. The gas, mostly oxygen, is secreted into the bladder by special glands, rendering the fish more buoyant; the gas is absorbed into the bloodstream by another special organ, reducing the overall buoyancy and allowing the fish to sink. Some deep-sea fishes may have oils, rather than gas, in the bladder. Other deep-sea and some bottom-living forms have much-reduced swim bladders or have lost the organ entirely.

The swim bladder of fishes follows the same developmental pattern as the lungs of land vertebrates. There is no doubt that the two structures have the same historical origin in primitive fishes. More or less intermediate forms still survive among the more primitive types of fishes, such as the lungfishes Lepidosiren and Protopterus.

The Circulatory System

The circulatory, or blood vascular, system consists of the heart, the arteries, the capillaries, and the veins. It is in the capillaries that the interchange of oxygen, carbon dioxide, nutrients, and other substances such as hormones and waste products takes place. The capillaries lead to the veins, which return the venous blood with its waste products to the heart, kidneys, and gills. There are two kinds of capillary beds: those in the gills and those in the rest of the body. The heart, a folded continuous muscular tube with three or four saclike enlargements, undergoes rhythmic contractions and receives venous blood in a sinus venosus. It passes the blood to an auricle and then into a thick muscular pump, the ventricle. From the ventricle the blood goes to a bulbous structure at the base of a ventral aorta just below the gills. The blood passes to the afferent (receiving) arteries of the gill arches and then to the gill capillaries. Their waste gases are given off to the environment, and oxygen is absorbed. The oxygenated blood enters efferent (exuant) arteries of the gill arches and then flows into the dorsal aorta. From there blood is distributed to the tissues and organs of the body. One-way valves prevent backflow. The circulation of fishes thus differs from that of the reptiles, birds, and mammals in that oxygenated blood is not returned to the heart prior to distribution to the other parts of the body.

Excretory Organs

The primary excretory organ in fishes, as in other vertebrates, is the kidney. In fishes some excretion also takes place in the digestive tract, skin, and especially the gills (where ammonia is given off). Compared with land vertebrates, fishes have a special problem in maintaining their internal environment at a constant concentration of water and dissolved substances, such as salts. Proper balance of the internal environment (homeostasis) of a fish is in a great part maintained by the excretory system, especially the kidney.

The kidney, gills, and skin play an important role in maintaining a fish's internal environment and checking the effects of osmosis. Marine fishes live in an environment in which the water around them has a greater concentration of salts than they can have inside their body and still maintain life. Freshwater fishes, on the other hand, live in water with a much lower concentration of salts than they require inside their bodies. Osmosis tends to promote the loss of water from the body of a marine fish and absorption of water by that of a freshwater fish. Mucus in the skin tends to slow the process but is not a sufficient barrier to prevent the movement of fluids through the permeable skin. When solutions on two sides of a permeable membrane have different concentrations of dissolved substances, water will pass through the membrane into the more concentrated solution, while the dissolved chemicals move into the area of lower concentration (diffusion).

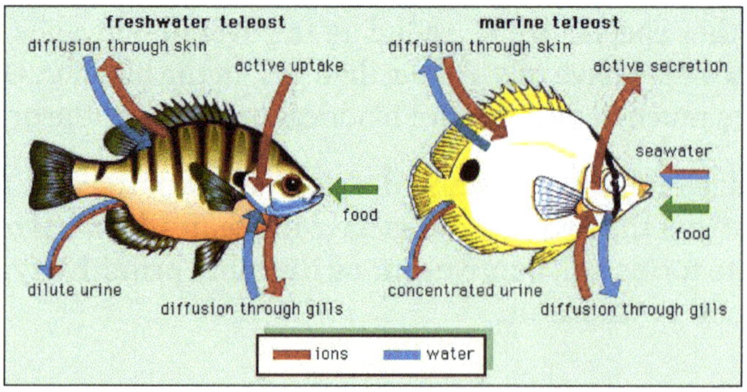

Generalized osmotic regulation in freshwater and marine teleost fishes.

The kidney of freshwater fishes is often larger in relation to body weight than that of marine fishes. In both groups the kidney excretes wastes from the body, but the kidney of freshwater fishes also excretes large amounts of water, counteracting the water absorbed through the skin. Freshwater fishes tend to lose salt to the environment and must replace it. They get some salt from their food, but the gills and skin inside the mouth actively absorb salt from water passed through the mouth.

This absorption is performed by special cells capable of moving salts against the diffusion gradient. Freshwater fishes drink very little water and take in little water with their food.

Marine fishes must conserve water, and therefore their kidneys excrete little water. To maintain their water balance, marine fishes drink large quantities of seawater, retaining most of the water and excreting the salt. Most nitrogenous waste in marine fishes appears to be secreted by the gills as ammonia. Marine fishes can excrete salt by clusters of special cells (chloride cells) in the gills.

There are several teleosts—for example, the salmon—that travel between fresh water and seawater and must adjust to the reversal of osmotic gradients. They adjust their physiological processes by spending time (often surprisingly little time) in the intermediate brackish environment.

Marine hagfishes, sharks, and rays have osmotic concentrations in their blood about equal to that of seawater and so do not have to drink water nor perform much physiological work to maintain their osmotic balance. In sharks and rays the osmotic concentration is kept high by retention of urea in the blood. Freshwater sharks have a lowered concentration of urea in the blood.

Endocrine Glands

Endocrine glands secrete their products into the bloodstream and body tissues and, along with the central nervous system, control and regulate many kinds of body functions. Cyclostomes have a well-developed endocrine system, and presumably it was well developed in the early Agnatha, ancestral to modern fishes. Although the endocrine system in fishes is similar to that of higher vertebrates, there are numerous differences in detail. The pituitary, the thyroid, the suprarenals, the adrenals, the pancreatic islets, the sex glands (ovaries and testes), the inner wall of the intestine, and the bodies of the ultimobranchial gland make up the endocrine system in fishes. There are some others whose function is not well understood. These organs regulate sexual activity and reproduction, growth, osmotic pressure, general metabolic activities such as the storage of fat and the utilization of foodstuffs, blood pressure, and certain aspects of skin colour. Many of these activities are also controlled in part by the central nervous system, which works with the endocrine system in maintaining the life of a fish. Some parts of the endocrine system are developmentally, and undoubtedly evolutionarily, derived from the nervous system.

The Nervous System and Sensory Organs

As in all vertebrates, the nervous system of fishes is the primary mechanism coordinating body activities, as well as integrating these activities in the appropriate manner with stimuli from the environment. The central nervous system, consisting of the brain and spinal cord, is the primary integrating mechanism. The peripheral nervous system, consisting of nerves that connect the brain and spinal cord to various body organs, carries sensory information from special receptor organs such as the eyes, internal ears, nares (sense of smell), taste glands, and others to the integrating centres of the brain and spinal cord. The peripheral nervous system also carries information via different nerve cells from the integrating centres of the brain and spinal cord. This coded information is carried to the various organs and body systems, such as the skeletal muscular system, for appropriate action in response to the original external or internal stimulus. Another branch of the nervous system, the autonomic nervous system, helps to coordinate the activities of many glands and organs and is itself closely connected to the integrating centres of the brain.

The brain of the fish is divided into several anatomical and functional parts, all closely interconnected but each serving as the primary centre of integrating particular kinds of responses and activities. Several of these centres or parts are primarily associated with one type of sensory perception, such as sight, hearing, or smell (olfaction).

Olfaction

The sense of smell is important in almost all fishes. Certain eels with tiny eyes depend mostly on smell for location of food. The olfactory, or nasal, organ of fishes is located on the dorsal surface of the snout. The lining of the nasal organ has special sensory cells that perceive chemicals dissolved in the water, such as substances from food material, and send sensory information to the brain by way of the first cranial nerve. Odour also serves as an alarm system. Many fishes, especially various species of freshwater minnows, react with alarm to a chemical released from the skin of an injured member of their own species.

Taste

Many fishes have a well-developed sense of taste, and tiny pitlike taste buds or organs are located not only within their mouth cavities but also over their heads and parts of their body. Catfishes, which often have poor vision, have barbels ("whiskers") that serve as supplementary taste organs, those around the mouth being actively used to search out food on the bottom. Some species of naturally blind cave fishes are especially well supplied with taste buds, which often cover most of their body surface.

Sight

Sight is extremely important in most fishes. The eye of a fish is basically like that of all other vertebrates, but the eyes of fishes are extremely varied in structure and adaptation. In general, fishes living in dark and dim water habitats have large eyes, unless they have specialized in some compensatory way so that another sense (such as smell) is dominant, in which case the eyes will often be reduced. Fishes living in brightly lighted shallow waters often will have relatively small but efficient eyes. Cyclostomes have somewhat less elaborate eyes than other fishes, with skin stretched over the eyeball perhaps making their vision somewhat less effective. Most fishes have a spherical lens and accommodate their vision to far or near subjects by moving the lens within the eyeball. A few sharks accommodate by changing the shape of the lens, as in land vertebrates. Those fishes that are heavily dependent upon the eyes have especially strong muscles for accommodation. Most fishes see well, despite the restrictions imposed by frequent turbidity of the water and by light refraction.

Pacific foureyed fish: Pacific foureyed fish (*Anableps dowei*).

Fossil evidence suggests that colour vision evolved in fishes more than 300 million years ago, but not all living fishes have retained this ability. Experimental evidence indicates that many shallow-water fishes, if not all, have colour vision and see some colours especially well, but some bottom-dwelling shore fishes live in areas where the water is sufficiently deep to filter out most if not all colours, and these fishes apparently never see colours. When tested in shallow water, they apparently are unable to respond to colour differences.

Hearing

Sound perception and balance are intimately associated senses in a fish. The organs of hearing are entirely internal, located within the skull, on each side of the brain and somewhat behind the eyes. Sound waves, especially those of low frequencies, travel readily through water and impinge directly upon the bones and fluids of the head and body, to be transmitted to the hearing organs. Fishes readily respond to sound; for example, a trout conditioned to escape by the approach of fishermen will take flight upon perceiving footsteps on a stream bank even if it cannot see a fisherman. Compared with humans, however, the range of sound frequencies heard by fishes is greatly restricted. Many fishes communicate with each other by producing sounds in their swim bladders, in their throats by rasping their teeth, and in other ways.

Other Senses (Touch, Pain and Special Senses)

A fish or other vertebrate seldom has to rely on a single type of sensory information to determine the nature of the environment around it. A catfish uses taste and touch when examining a food object with its oral barbels. Like most other animals, fishes have many touch receptors over their body surface. Pain and temperature receptors also are present in fishes and presumably produce the same kind of information to a fish as to humans. Fishes react in a negative fashion to stimuli that would be painful to human beings, suggesting that they feel a sensation of pain.

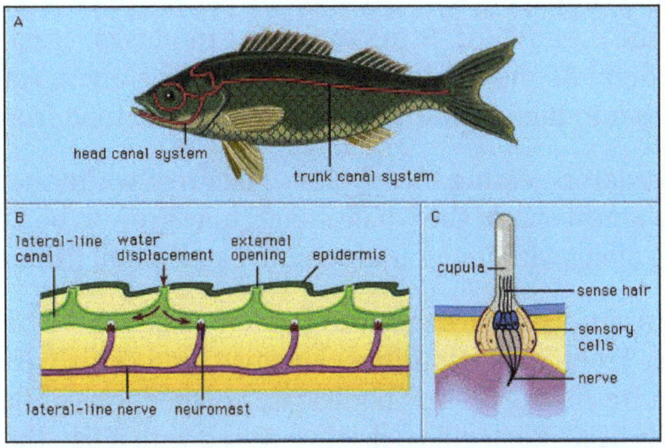

Lateral line system of a fish. (A) Bodily location of lateral lines.
(B) Longitudinal section of a canal. (C) Superficial neuromast.

An important sensory system in fishes that is absent in other vertebrates (except some amphibians) is the lateral line system. This consists of a series of heavily innervated small canals located in the skin and bone around the eyes, along the lower jaw, over the head, and down the mid-side of the body, where it is associated with the scales. Intermittently along these canals are located tiny sensory organs (pit organs) that apparently detect changes in pressure. The system allows a fish to sense changes in water currents and pressure, thereby helping the fish to orient itself to the various changes that occur in the physical environment.

Types of Fishes

They are divided into these groups because of the structure of their mouths and the types of skeletons they have. There are jawless fishes, cartilaginous fishes, and bony fishes. All three types are "cold-blooded," or ectotherms. This means that they can't control their body temperature internally. Their body temperature is similar to the temperature of their outside environment.

The jawless fishes were the earliest vertebrates. Today there are only about sixty species still living. These fish have no scales. Their skeletons are made of cartilage, a firm, flexible material like the end of your nose. They also do not have pairs of fins like most fish. But the most remarkable thing about these fish is that they do not have jaws! Instead, the mouths of these fishes have structures for scraping, stabbing, and sucking.

Hagfishes and lampreys are the only living kinds of jawless fishes. Hagfishes look like big slimy worms! They crawl into the bodies of dead or dying fish and use their sandpapery tongue to scrape the tissues.

Lampreys have a mouth like a suction cup with many tiny, but sharp, teeth. Many of them are parasites that feed on other fish. A parasite is an organism that lives off another organism without helping the host organism in any way. Lampreys attach their mouths to a healthy fish and then suck the tissues and blood out of the healthy fish. There are many lampreys in the Great Lakes.

The second type of fish is the cartilaginous fishes. These include sharks, rays, and skates. Their skeletons are made of cartilage, just like the skeletons of the jawless fishes. However, they do have jaws and pairs of fins. Their bodies are covered with pointed, tooth like scales. This gives their bodies a texture rougher than sandpaper. All cartilaginous fishes are carnivores. Rays, like sting rays and manta rays, live on the ocean floor. Sting rays hunt mollusks, crustaceans, and small fishes. Manta rays are filter-feeders. As they swim with their mouths open, water containing plankton, very small animals, comes into their mouths. They filter out the food from the water.

Most sharks are active predators, eating mostly fish. The great white shark also preys upon ocean mammals like seals. The largest shark, the whale shark, may grow to be forty feet long. It feeds like a filter-feeder, swimming along with its mouth open to catch small fish and plankton.

The third type is the bony fishes. These are the most familiar fish. Bony fish make up about ninety-five percent of all fish species. Trout, goldfish, tuna, clownfish, and catfish are all kinds of bony fishes. They live in both salt and fresh water. Their bodies are covered with scales. Their gills are inside a pocket on the sides of their head. Each gill pocket is covered with a flap that opens to release water.

Bony fish have scales that overlap like the shingles on a roof. Scales are made from a substance like your fingernails. Mucus covers and protects the scales. This slimy coating is important because it protects the fish from parasites, infections, and diseases. It helps wounds heal and even helps the fish move easily through the water. The paired fins help the fish stay balanced and upright in the water. Fins are used for swimming. Fins are thin membranes stretched across a framework of bones. Like a wide boat paddle, a fin provides a large surface to push against the water.

Most all fish breathe with gills. As a fish swims, it automatically opens its mouth. It continuously gulps water, even when it sleeps. The water, which contains oxygen, moves through openings in the fish's throat that lead to the gills. Gills, which look like tiny feathers, are red because of the many blood vessels in them. As water flows over the gills, oxygen moves from the water into the fish's blood. Carbon dioxide, a waste product, moves out of the blood into the water. After flowing over the gills, water leaves the fish through slits beneath the gill flaps.

From the gills, the blood travels through the fish's body. Blood is pumped from the heart to the gills, from the gills to the rest of the body, and back again to the heart. Each cell of the fish's body receives oxygen from the blood and releases its carbon dioxide into the blood. Interestingly, a fish's heart has only two chambers.

Fish come in all shapes, colors, and sizes. Scientists estimate that there are more than 20,000 different species of fish! Fish have lived on Earth before the dinosaurs. Fish have adapted to almost every watery environment and type of food.

Types of Pet Fish

Betta

For ease of maintenance, bettas can't be beat. Also known as Siamese fighting fish, bettas require separation from most other species, which means these brilliantly colored swimmers do well in small fish bowls. While keeping male bettas separate is imperative, some female bettas can live in tanks with other fish. Bettas are a cold-water species.

Goldfish

Another cold-water fish, goldfish belong to the carp family. Because they enjoy cool water temperatures, keep goldfish in a separate tank from warm water fish. Avoid keeping goldfish in a bowl, as they can grow quite long and need sufficient swimming room. Because they do grow so large, don't overcrowd your goldfish tank. Well-kept goldfish can live for many years.

Angelfish

Large, lovely and graceful, angelfish appear in various color patterns. Because of their size when full-grown, angelfish require at least a 55-gallon tank. Angelfish do well with other fish species (although they may eat very small fish) but can fight with each other. Provide plenty of plants in the aquarium, as angelfish like to hide beneath them.

Catfish

Catfish aren't the most spectacular fish in a tank, but they serve an important purpose. This low-maintenance, bottom-dwelling fish consume algae growing in the tank, so they aid in keeping the tank clean. Choose among various catfish species and colors. Most types of catfish are compatible with the fish commonly kept in community tanks.

Guppies

These easy-care aquarium fish appear in a variety of colors. There is one drawback to guppies: They breed constantly, so if you have male and females together, the offspring can soon overwhelm a tank. For best results, choose all males or all females. The former sport longer tails and brighter colors.

Mollies

These small, hardy fish do best in tanks with heavy filtration. Unlike many tropical fish breeds, mollies bear live young rather than lay eggs. If you want to raise mollies, a ratio of one male per three to four females works best. Too many male mollies stress out the females with constant breeding. Mollies tend to nip the fins of other species, so you may want to keep them in a separate aquarium. Keeping a few mollies in a community tank can work out well.

Neon Tetras

These small, translucent blue and red fish hail from the Amazon jungles. Because they aren't aggressive, they are a welcome and colorful addition to any freshwater tank. Keep neon tetras in a school of at least six fish, and preferably more. You'll find them swimming in the middle of your tank.

Platies

Platies are ideal for those getting started in tropical fishkeeping. They are available in dozens of shades, and their harmonious natures mean they are good community fish. If you only have room for a smaller tank, around 10 gallons, little platies make a good choice. These active fish tend to swim mid-tank.

Swordtails

The larger the tank, the larger you can expect swordtails to grow. The name comes from their sword-like tails, and these fish most often sport a red body with a black tail. Swordtails make good fish for the beginning hobbyist, as they don't require any special care, although males may fight with each other. It's best to keep swordtails in a larger tank, or at least one larger than 10 gallons.

Zebra Danios

Named for their striped bodies, zebra danios are tough fish. They can thrive in a variety of water temperatures, even into the low 60s. Unlike many species, zebra danios mate for life. These active fish are perhaps the easiest of the egg-laying species, if you want to breed them. Zebra danios swim all over the tank, and make good community fish due to their peaceful nature.

Fish Development

In recent years, the teleost fish *Danio rerio,* known as the zebrafish, has become a favorite organism of those who wish to study vertebrate development. Zebrafish have large broods, breed all year, are easily maintained, have transparent embryos that develop outside the mother (an important feature for microscopy), and can be raised so that mutants can be readily screened and propagated. In addition, they develop rapidly, so that at 24 hours after fertilization, the embryo has formed most of its tissues and organ primordia and display the characteristic tadpole-like form. Therefore, much of the description of fish development below is based on studies of this species.

Cleavage in Fish Eggs

In fish eggs, cleavage occurs only in the blastodisc, a thin region of yolk-free cytoplasm at the animal cap of the egg. Most of the egg cell is full of yolk. The cell divisions do not completely divide the egg, so this type of cleavage is called meroblastic. Since only the cytoplasm of the blastodisc becomes the embryo, this type of meroblastic cleavage is called discoidal. Scanning electron micrographs show beautifully the incomplete nature of discoidal meroblastic cleavage in fish eggs. The calcium waves initiated at fertilization stimulate the contraction of the actin cytoskeleton to squeeze non-yolky cytoplasm into the animal pole of the egg. This converts the spherical egg into a more pear-shaped structure, with an apical blastodisc. Early cleavage divisions follow a highly reproducible pattern of meridional and equatorial cleavages. These divisions are rapid, taking about 15 minutes each. The first 12 divisions occur synchronously, forming a mound of cells that sits at the animal pole of a large yolk cell. These cells constitute the blastoderm. Initially, all the cells maintain some open connection with one another and with the underlying yolk cell so that moderately sized (17-kDa) molecules can pass freely from one blastomere to the next.

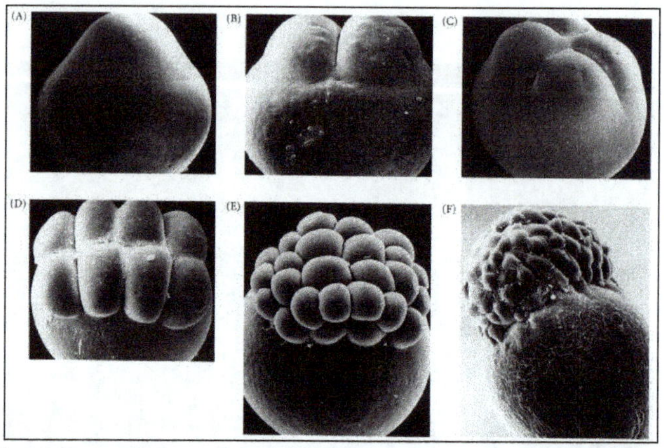

Discoidal cleavage in a zebrafish egg. (A) 1-cell embryo. The mound atop the cytoplasm is the blastodisc region. (B) 2-cell embryo. (C) 4-cell embryo. (D) 8-cell embryo, wherein two rows of four cells are formed. (E) 32-cell embryo. (F) 64-cell embryo, wherein the blastodisc can be seen atop the yolk cell.

Beginning at about the tenth cell division, the onset of the midblastula transition can be detected: zygotic gene transcription begins, cell divisions slow, and cell movement becomes evident. At this time, three distinct cell populations can be distinguished. The first of these is the yolk syncytial layer (YSL). The YSL is formed at the ninth or tenth cell cycle, when the cells at the vegetal edge of the blastoderm fuse with the underlying yolk cell. This fusion produces a ring of nuclei within the part of the yolk cell cytoplasm that sits just beneath the blastoderm. Later, as the blastoderm expands vegetally to surround the yolk cell, some of the yolk syncytial nuclei will move under the blastoderm to form the internal YSL, and some of the nuclei will move vegetally, staying ahead of the blastoderm margin, to form external YSL. The YSL will be important for directing some of the cell movements of gastrulation. The second cell population distinguished at the midblastula transition is the enveloping layer. It is made up of the most superficial cells of the blastoderm, which form an epithelial sheet a single cell layer thick. The EVL eventually becomes the periderm, an extraembryonic protective covering that is sloughed off during later development.

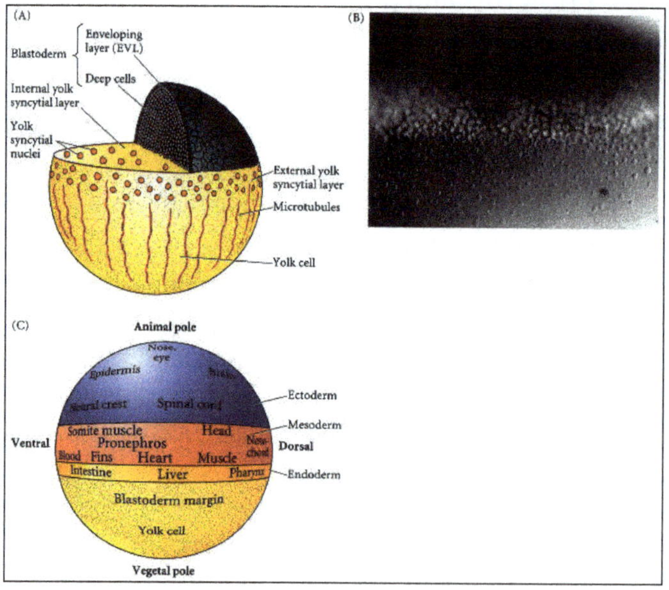

Figure shows fish blastula. (A) Prior to gastrulation, the deep cells are surrounded by the EVL.

The animal surface of the yolk cell is flat and contains the nuclei of the YSL. Microtubules extend through the yolky cytoplasm and through the external region of the YSL. (B) Late-blastula stage embryo of the minnow *Fundulus,* showing the external YSL. The nuclei of these cells were derived from cells at the margin of the blastoderm, which released their nuclei into the yolky cytoplasm. (C) Fate map of the deep cells after cell mixing has stopped. The lateral view is shown, and not all organ fates are labeled (for the sake of clarity).

Between the EVL and the YSL are the deep cells. These are the cells that give rise to the embryo proper. The fates of the early blastoderm cells are not determined, and cell lineage studies (in which a nondiffusible fluorescent dye is injected into one of the cells so that the descendants of that cell can be followed) show that there is much cell mixing during cleavage. Moreover, any one of these cells can give rise to an unpredictable variety of tissue descendants. The fate of the blastoderm cells appears to be fixed shortly before gastrulation begins. At this time, cells in specific regions of the embryo give rise to certain tissues in a highly predictable manner, allowing a fate map to be made.

Gastrulation in Fish Embryos

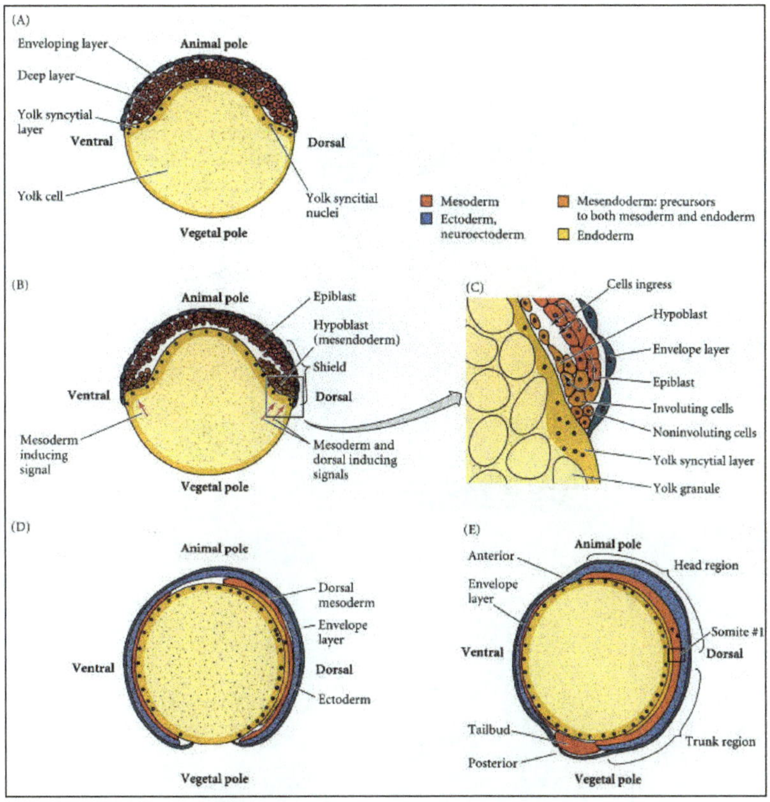

Figure shows cell movements during gastrulation of the zebrafish. (A) The blastoderm at 30% completion of epiboly (about 4.7 hours). (B) Formation of the hypoblast, either by involution of cells at the margin of the epibolizing blastoderm or by delamination of cells from the epiblast (6 hours). (C) Close-up of the marginal region. (D) At 90% epiboly (9 hours), mesoderm can be seen surrounding the yolk, between the endoderm and ectoderm. (E) Completion of gastrulation (10.3 hours).

The first cell movement of fish gastrulation is the epiboly of the blastoderm cells over the yolk. In the initial phase, the deep blastoderm cells move outwardly to intercalate with the more superficial cells. Later, these cells move over the surface of the yolk to envelop it completely. This movement is not due to the active crawling of the blastomeres, however. Rather, the movement is provided by the autonomously expanding YSL "within" the animal pole yolk cytoplasm. The EVL is tightly joined to the YSL and is dragged along with it. The deep cells of the blastoderm then fill in the space between the YSL and the EVL as epiboly proceeds. This can be demonstrated by severing the attachments between the YSL and the EVL. When this is done, the EVL and deep cells spring back to the top of the yolk, while the YSL continues its expansion around the yolk cell. The expansion of the YSL depends on a network of microtubules in the YSL, and radiation or drugs that block the polymerization of tubulin inhibit epiboly.

During migration, one side of the blastoderm becomes noticeably thicker than the other. Cell-labeling experiments indicate that the thicker side marks the site of the future dorsal surface of the embryo.

The Formation of Germ Layers

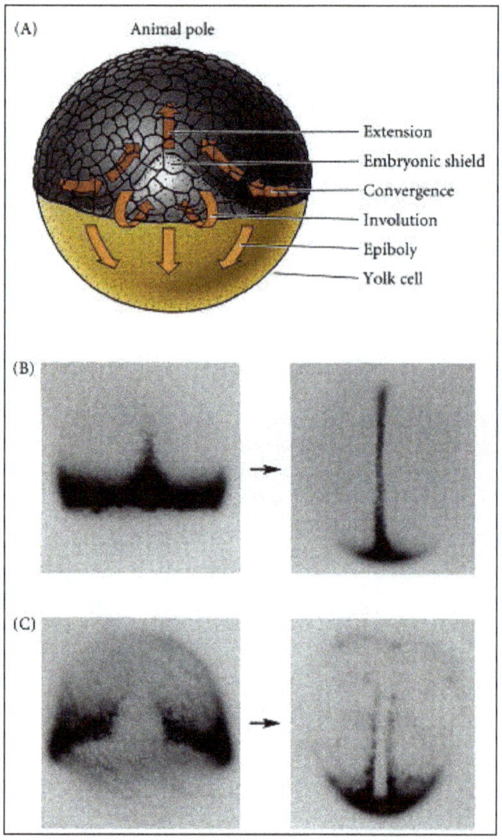

Figure shows convergence and extension in the zebrafish gastrula. (A) Dorsal view of the convergence and extension movements during zebrafish gastrulation. Epiboly spreads the blastoderm over the yolk; involution or ingression generates the hypoblast; convergence and extension bring the hypoblast and epiblast cells to the dorsal side to form the embryonic shield. Within the shield, intercalation extends the chordamesoderm toward the animal pole. (B) Convergent extension of chordamesoderm is shown by those cells expressing the gene *no tail,* a gene that is expressed by notochord cells. (C) Convergent extension of paraxial mesodermal cells (marked by their expression of the *snail* gene) to flank the notochord.

After the blastoderm cells have covered about half the zebrafish yolk cell (and earlier in fish eggs with larger yolks), a thickening occurs throughout the margin of the epibolizing blastoderm. This thickening is called the germ ring, and it is composed of a superficial layer, the epiblast, and an inner layer, the hypoblast. We do not understand how the hypoblast is made. Some research groups claim that the hypoblast is formed by the *involution* of superficial cells under the margin followed by their migration toward the animal pole. The involution begins at the future dorsal portion of the embryo, but occurs all around the margin. Other laboratories claim that these superficial cells *ingress* to form the hypoblast. It is possible that both mechanisms are at work, with different modes of hypoblast formation predominating in different species. Once formed, however, the cells of both the epiblast and hypoblast intercalate on the future dorsal side of the embryo to form a localized thickening, the embryonic shield. this shield is functionally equivalent to the dorsal blastopore lip of amphibians, since it can organize a secondary embryonic axis when transplanted to a host embryo. Thus, as the cells undergo epiboly around the yolk, they are also involuting at the margins and converging anteriorly and dorsally toward the embryonic shield. The hypoblast cells of the embryonic shield converge and extend anteriorly, eventually narrowing along the dorsal midline of the hypoblast. This movement forms the chordamesoderm, the precursor of the notochord. The cells adjacent to the chordamesoderm, the paraxial mesoderm cells, are the precursors of the mesodermal somites. The concomitant convergence and extension in the epiblast brings the presumptive neural cells from all over the epiblast into the dorsal midline, where they form the neural keel. The rest of the epiblast becomes the skin of the fish. The zebrafish fate map, then, is not much different from that of the frog or other vertebrates. If one conceptually opens a *Xenopus* blastula at the vegetal pole and stretches the opening into a marginal ring, the resulting fate map closely resembles that of the zebrafish embryo at the stage when half of the yolk has been covered by the blastoderm.

Axis Formation in Fish Embryos

Dorsal-ventral Axis Formation: The Embryonic Shield

The embryonic shield is critical in establishing the dorsal-ventral axis in fishes. It can convert lateral and ventral mesoderm (blood and connective tissue precursors) into dorsal mesoderm (notochord and somites), and it can cause the ectoderm to become neural rather than epidermal. This was shown by transplantation experiments in which the embryonic shield of one early-gastrula embryo was transplanted to the ventral side of another. Two axes formed, sharing a common yolk cell. Although the prechordal plate and the notochord were derived from the donor embryonic shield, the other organs of the secondary axis came from host tissues that would have formed ventral structures. The new axis has been induced by the donor cells. In the embryo that had had its embryonic shield removed, no dorsal structures formed, and the embryo lacked a nervous system. These experiments are similar to those performed on amphibian gastrulae by Spemann and Mangold, and they demonstrate that the embryonic shield is the homologue of the dorsal blastopore lip, the amphibian organizer.

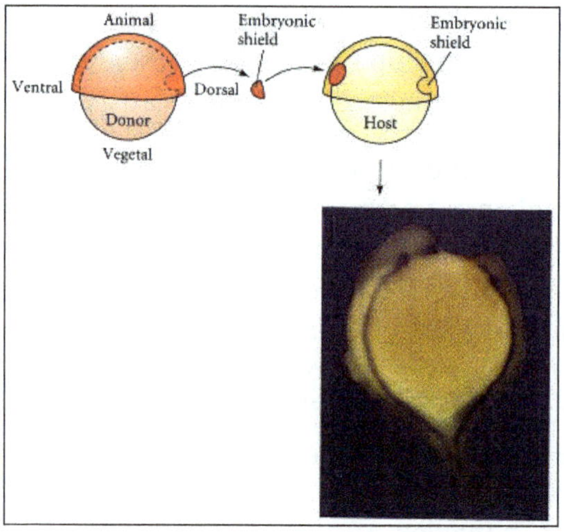

Figure shows the embryonic shield as organizer in the fish embryo. A donor embryonic shield (about 100 cells from a stained embryo) is transplanted into a host embryo at the same early-gastrula stage. The result is two embryonic axes joined to the host's yolk cell. In the photograph, both axes have been stained for *sonic hedgehog* mRNA, which is expressed in the ventral midline. (The embryo to the right is the secondary axis.)

Like the amphibian dorsal blastopore lip, the embryonic shield forms the prechordal plate and the notochord of the developing embryo. The precursors of these two regions are responsible for inducing the ectoderm to become neural ectoderm. Moreover, the presumptive notochord and prechordal plate appear to do this in a manner very much like that of their homologous structures in amphibians. In both fishes and amphibians, BMP proteins made in the ventral and lateral regions of the embryo would normally cause the ectoderm to become epidermis. The notochord of both fishes and amphibians secretes factors that block this induction and thereby allow the ectoderm to become neural. In fishes, the BMP that ventralizes the embryo is BMP2B. The protein secreted by the chordamesoderm that binds with and inactivates BMP2B is a chordin-like paracrine factor called Chordino. If the *chordino* gene is mutated, the neural tube fails to form. It is hypothesized that different concentrations of BMP2B pattern the ventral and lateral regions of the zebrafish ectoderm and mesoderm, and that the ratio between Chordino and BMP2B may specify the position along the dorsal-ventral axis. In fishes, however, the notochord may not be the only structure capable of producing the proteins that block BMP2B. If the notochord fails to form (as in the *floating head* or *no tail* mutations), the neural tube will still be produced. It is possible that the notochordal precursor cells (which are produced in these mutations) are still able to induce the neural tube, or that the dorsal portion of the somite precursors can compensate for the lack of a notochord.

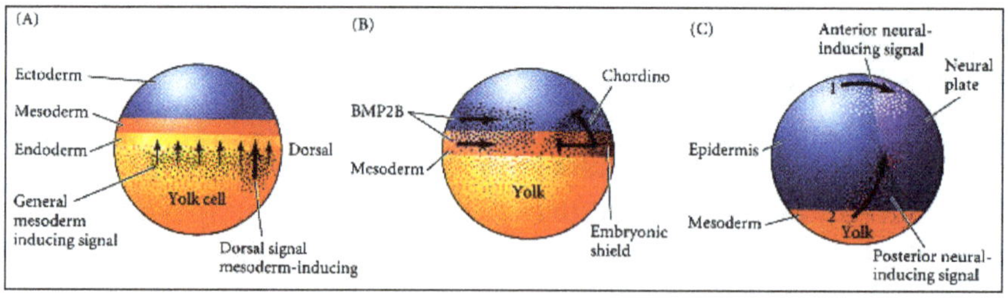

Figure shows axis formation in the zebrafish embryo. (A) Prior to gastrulation, the zebrafish blastoderm is arranged with the presumptive ectoderm near the animal pole, the presumptive mesoderm beneath it, and the presumptive endoderm sitting atop the yolk cell. The yolk syncitial layer (and possibly the endoderm) sends two signals to the presumptive mesoderm. One signal (lighter arrows) induces the mesoderm, while a second signal (heavy arrow) specifically induces an area of mesoderm to become the dorsal mesoderm (embryonic shield). (B) Formation of the dorsal-ventral axis. During gastrulation, the ventral mesoderm secretes BMP2B (arrows) to induce the ventral and lateral mesodermal and epidermal differentiation. The dorsal mesoderm secretes factors (such as Chordino) that block BMP2B and dorsalize the mesoderm and ectoderm (converting the latter into neural tissue). (C) Recent studies have identified two signaling centers for anterior-posterior polarity, one (1) at the border of the neural and non-neural ectoderm, which induces anterior neural cell types, and the other (2) at the lateral margin, which generates a posteriorizing signal.

The embryonic shield appears to acquire its organizing ability in much the same way as its amphibian counterparts. In amphibians, the endoderm cells beneath the dorsal blastopore lip (i.e., the Nieuwkoop center) accumulate β-catenin. This protein is critical in amphibians for the ability of the endoderm to induce the cells above them to become the dorsal lip (organizer) cells. In zebrafish, the nuclei in that part of the yolk syncytial layer that lies beneath the cells that will become the embryonic shield similarly accumulate β-catenin. This protein distinguishes the dorsal YSL from the lateral and ventral YSL regions. Inducing β-catenin accumulation on the ventral side of the egg causes dorsalization and a second embryonic axis. In addition, just prior to gastrulation, the cells of the dorsal blastopore margin synthesize and secrete Nodal-related proteins. These induce the precursors of the notochord and prechordal plate to activate *goosecoid* and other genes. Thus, the embryonic shield is considered equivalent to the amphibian organizer, and the dorsal part of the yolk cell can be thought of as the Nieuwkoop center of the fish embryo.

Nuclear localization of β-catenin marks the dorsal side of the *Xenopus* blastula (the larger one) and helps form its Nieuwkoop center beneath the organizer. In the zebrafish late blastula, nuclear localization of β-catenin is seen in the yolk syncytial layer nuclei beneath the future embryonic shield.

Anterior-posterior Axis Formation: Two Signaling Centers

As is evident from, when a second dorsal-ventral axis is experimentally induced in zebrafish eggs, both the regular and the induced axes have the same anterior-posterior polarity. Both heads are at the former animal cap, and both tails are located vegetally. Indeed, the anterior-posterior axis is specified during oogenesis, and the animal cap marks the anterior of the embryo. This axis becomes stabilized during gastrulation through two distinct signaling centers. First, a small group of anterior neural cells at the border between the neural and surface ectoderm (a region that become the pituitary gland, nasal placode, and anterior forebrain) secrete compounds that cause anterior development. If these anterior neural cells are experimentally placed more posteriorly in the embryo, they will cause the neural cells near them to assume the characteristics of forebrain neurons. The second signaling center, in the posterior of the embryo, consists of lateral mesendoderm precursors at the margin of the gastrulating blastoderm. These cells produce caudalizing compounds, most likely Nodal-related proteins and activin. If transplanted adjacent to anterior neural ectoderm, this tissue will transform the presumptive forebrain tissue into hindbrain-like structures.

Left-right Axis Formation

Little is known about left-right axis formation in fishes. In zebrafish, a gene called *nodal-related-2* is expressed solely in the left side of the lateral plate mesoderm. Nodal-related-2 is a paracrine factor of the TGF-β family.

Fish Habitats

Habitat determines which fish species live in various bodies of water. Trout live in lakes with clear, cold and deep water. Most lakes, by comparison, have cool, moderately clear, moderately deep water, where bass, pike, walleyes or panfish may live. Warm, shallow lakes are bullhead and carp territory.

In rivers, trout live in the youngest, cleanest and coldest portions. These areas usually are nearer the source of a river, well upstream from the river mouth. Follow a river downstream, and the fish species often change from trout to bass and walleyes, and finally, to catfish and carp. Obviously, where you choose to fish determines which fish species are available.

Fish like bass and panfish also thrive in ponds, often along with catfish. Occasionally, ponds are home to trout. "Pay ponds" are popular in some areas. Fish are stocked into these ponds and anglers pay to fish.

Gulf and Atlantic Coast waters of the South offer redfish and speckled trout, along with sheepshead and sometimes sharks and other larger predators. Stripers and shad are available farther north along the Atlantic Coast. And along the West Coast, fish like sand bass and croaker predominate, along with salmon and halibut farther north.

Fish Migration

Many species of fish wander annually through a particular area of the ocean. Some are true migrants, travelling regularly over great distances. Young fish usually leave the spawning grounds for areas where they develop into juveniles, before joining the adult stock at the feeding grounds. Adults move to the spawning grounds, then return to the feeding grounds. Migratory patterns of fish are related to oceanographic factors and to currents. Eggs, larvae, and young drift passively with the current, although migration of adult fish toward breeding grounds is usually against the current. Adult movements thus are directional rather than passive, and the fish respond to environmental conditions—e.g., climate.

Three categories of migratory fishes can be distinguished: Oceanodromous, anadromous, and catadromous.

Oceanodromous Fish

Oceanodromous fish, which occur widely throughout the world's oceans, live and migrate wholly in the sea. They differ mainly from one another by the method and extent of their migration.

Herring (*Clupea harengus*), extensively studied because of their economic importance, are the best known of the oceanodromous type and can be classified into several populations, or local races, which do not mix freely. In addition, each has a particular migratory behaviour. In the North Sea, distinct groups spawn in different seasons and on different grounds: Buchan herring spawn in August and September off the coast of Scotland and migrate to the coast of southwestern Norway; Dogger Bank herring spawn in September and October in the central part of the North Sea and along the English coast and then migrate to the Skagerrak, an arm of the North Sea between Denmark and Norway; Downs herring spawn from November to January off the French coast, mainly between Dunkirk and Fécamp, then feed in summer in the middle and northern parts of the North Sea, sharing the feeding grounds with other populations. The diversity of migration and of reproductive seasons is closely connected with the annual cycle of oceanographic conditions in the North Sea.

Cod (*Gadus morhua*) have migration patterns similar to those of herring. The migrations of other fish cover even greater distances; in the Atlantic, for example, white tuna (*Germo alalunga*) are found in winter around the Azores and the Canary Islands, where they spawn in spring. They then migrate northward to the Gulf of Gascogne and afterward to the waters around Iceland, arriving there in July. Populations of red tuna (*Thunnus thynnus*) occur throughout the Mediterranean Sea and the eastern Atlantic. In May and June they spawn in the western Mediterranean. During summer they spread northward to feed, finally reaching the Arctic Ocean. Similar migrations occur along the North American coast in the Atlantic and throughout the Pacific.

Anadromous Fish

Anadromous fish live in the sea and migrate to fresh water to breed. Their adaptations to conditions of different habitats are precise, particularly with regard to salinity of the water.

Salmon (*Salmo, Oncorhynchus*) spawn in the cold, clear waters of lakes or upper streams. Eggs are laid in gravel beds. The young of the Atlantic salmon remain in fresh water for two to three years, sometimes as long as six; Pacific salmon sometimes migrate to the sea in their first year. Adult fish usually remain in the sea for two or three winters, but sometimes only one. Then, as grilse (adolescents) or as adults, they return to fresh water and spawn, after changes occur in colour and other external features. Some Atlantic salmon die in fresh water after a single spawning; others return to the sea.

Migration: sockeye salmon. Sockeye salmon (*Oncorhynchus nerka*) leaping the waterfalls of the Brooks River in order to spawn upstream, Katmai National Park and Preserve, Alaska.

The tagging of salmon has shown that European types may cross from Norway to Scotland, as well as the reverse. Pacific salmon are probably distributed over the Pacific Ocean and Bering Sea, between latitudes 45 °N and 65 °N with surface waters of 2 to 11 °C (36 to 52 °F).

Experiments in Canada and the United States, in which young salmon migrating to the Pacific have been tagged, have shown that a high proportion of the fish return to the river where they hatched. Tagging of Atlantic salmon has shown that a few survivors have migrated two or even three times to a particular river in successive years. Adults reared from experimentally transplanted eggs return to the stream where they were hatched or grew, not to the stream where the eggs were laid. Aside from other means of orientation, such as reference to celestial features, topographical features are believed to play an important part in recognition of the original habitat. The sense of smell, or olfaction, however, has the most important role. Experiments have shown that migrating salmon are attracted to the waters of the stream in which they are going to spawn. Experiential imprinting at an early stage of development enables a grown fish to respond to waters that contain substances with a particular odour or that have a characteristic temperature.

Catadromous Fish

Catadromous fish spend most of their lives in fresh water, and then migrate to the sea to breed. This type is exemplified by eels of the genus *Anguilla*, numbering 16 species, the best-known of which are the North American eel (*A. rostrata*) and the European eel (*A. anguilla*).

European eels and North American eels spawn in warm saline waters of the Atlantic, at depths of

400 to 700 metres (about 1,300 to 2,300 feet), in an area centred near latitude 26 °N longitude 55 °W called the Sargasso Sea. The pelagic eggs develop into leptocephali—transparent, leaflike forms with relatively small heads—that are carried by the Gulf Stream to the shallow waters of the continental shelves. When they are about two and one-half years old and about eight centimetres long (a little more than three inches), a metamorphosis occurs. The leptocephali are transformed into so-called elvers, which are bottom-dwelling, pigmented, and cylindrical in form. They arrive in coastal waters as glass eels and begin to swim upstream in freshwater streams in spring. Their migration upstream is spectacular, as the young fish gather by millions, forming a dense mass several miles long. In freshwater the eels grow to full size, becoming yellow eels. They live as such for 10 to 15 years before changing into silver eels, with enlarged eyes; they swim downstream to the sea, return to the spawning grounds (Sargasso Sea), and die. The migration of these eels is not entirely understood, particularly their return to the Sargasso Sea. It may be that European eels and North American eels belong to the same species.

Fish Aggression

Many species of fish are naturally aggressive which can lead to problems with attacks on other fish in the tank. That said, this isn't a reason to avoid these species, though, as many of them are fun, interesting, and intelligent fish that, given the right situation, can add a lot to an aquarium. It's important to remember certain things about aggressive fish, so that you can avoid any attacks or deaths.

Aggressive Fish like to Establish their Own Territory

An aggressive fish will attach itself to your aquarium or a section of your aquarium shortly after being introduced to its new home. It will subsequently defend this piece of territory against all encroachers. Don't worry, though, as there are ways you can help remedy this tendency.

- Before you add new fish, try to rearrange the landscape of your tank. This will help to nullify any territorial claims and leave all of the fish on equal footing.
- Make sure there are plenty of hiding places in your tank. More aggressive fish will like that they have a defined space to protect. Less aggressive fish will appreciate the opportunity to stay out of sight of the more dominant tank-mates.
- If you are going to have more than one dominant or aggressive fish, try to add them at the same time. This way they will be in equal position to stake a claim to a piece of land.
- Be sure not to overstock your tank. In a crowded environment, an aggressive fish could get out of control. Try to research all of the species that you plan on having to ensure proper tank size.

Aggressive Fish will Fight for Food

As with territory, many aggressive fish will compete for food and will fight off fish they perceive to threaten their meals. Make sure you spread food evenly throughout the tank when feeding and try to offer a variety of food.

Aggressive Fish are More Likely to Fight with other Aggressive Fish

Since many aggressive fish perceive themselves to be dominant over their tank, they will fight anyone who threatens this status. Thus, you should try to avoid having two dominant fish in a tank. If you choose to put two together, make sure you add them at the same time, or place a mirror next to the aquarium to distract them from each other. Additionally, consider buying two aggressive fish from the same aquarium, they will have already established their relationship and will be less likely to compete.

Size Differences Exacerbate Aggression

Larger fish tend to be more aggressive, and are especially troublesome to their smaller companions. When you are stocking your tank, try to pick fish that will grow to roughly the same size. This will help reduce the possibility of attacks.

Aggression can be a Warning Sign

Some fish become aggressive when they are sick, and others tend to be aggressive towards these sick fish. Whichever the case, when a fish is being aggressive, you should observe him to see if there may be an underlying cause. Helping alleviate that cause could be the key to reducing aggression.

Identifying the Signs of Aggression in Fish

If you see one of your fish chasing another fish around the tank, it is an obvious sign of aggression but not all signs of aggression are quite so obvious. In most cases, fights among aquarium fish happen when you are not looking – you just discover the damage after the fight is over. This being the case, you may need to look for signs of fighting rather than watching for the actual signs of aggression. Things like split fins, changes in behavior among your fish, changes in territory, scratches and scrapes, missing scales, and obvious wounds are all signs of fighting.

In addition to learning to identify the signs of aggression in aquarium fish, you should also know what triggers cause aquarium fish to act aggressively. The most common causes of fights among aquarium fish are food, mates, and territory. Territory is the biggest problem when it comes to aquarium fish aggression, especially if you have stocked your tank with very large fish. In most cases, aggressive aquarium fish will establish their territory around some kind of hiding place like a rock cave or a piece of driftwood – they guard their territory fiercely and punish any fish that comes too close.

Aside from territory battles, aquarium fish aggression is also frequently caused by a fight for resources

– this includes both food and mates. Female fish do not tend to be as aggressive about fighting for mates (that is the male's job) but they can become extremely protective and territorial when it comes to guarding their eggs or fry. Male fish will sometimes fight to the death over a female, especially if there is more than one male of the same species in the tank and not enough females to go around. Male fish will also sometimes fish males of other species, especially if they have similar coloring or patterns.

Landscaping Techniques to Minimize Aggression

When it comes to decorating an aquarium for large and aggressive species there are several key things you need to incorporate. You need to include plenty of rocks and driftwood to provide individual territories as well as some plants and other decorations to break up sightlines. When choosing rockwork for your tank, keep in mind that jagged edges and rough materials can cause serious injury to your fish if they run into it while being chased by another fish. Smooth rocks, like river rocks, are generally the best choice for rockwork and you can also incorporate pieces of driftwood. Use these pieces smartly to not only incorporate territories into your tank, but to increase its aesthetic appeal as well.

In addition to thinking about what kind of rockwork and driftwood you want to use in your tank, you also need to think carefully about whether you are going to use life plants or not. Live plants can give your tank a pleasing natural appearance and they can help to oxygenate your tank water, but they are not always the best choice for a large tank housing aggressive species. Large fish can easily uproot aquarium plants, damaging the roots, or they can eat them. Either way, you need to think carefully before adding live plants to your tank. Your best bet might be to purchase high-quality synthetic plants that look real but will stand up to some rough handling.

When arranging the design elements in your tank you need to be conscious about creating separate territories while also providing extra hiding places for the other fish in your tank. Clay pots, rockwork caves, pieces of driftwood, and other decorations can be arranged in ways to create hiding places. If you are trying to build specific territories I your tank, be sure to include a hiding place (like a cave or clay pot) with some smaller rocks scattered about and some synthetic plants to provide cover. Be sure to include at least one territory per male in your tank – the females will generally swim freely between the territories as they like. If you find that the arrangement of territories in your tank isn't working, you can always rearrange them for better results.

Other Tips for Maintaining Aggressive Fish

Aside from thoughtfully aquascaping your aquarium to minimize aggression between your fish, there are also a few simple things you can do to keep the peace. For one thing, avoid lining your tank with a substrate bed that is too deep – about 1 to 2 inches is perfect. This may not seem like an adequate depth but you will find that when it comes time to vacuum your gravel during routine water changes that it is much easier to get up all the accumulated waste and debris from a 2-inch gravel bed than from a 4-inch bed. This tip goes hand-in-hand with a recommendation to change your tank water frequently. Many aggressive fish are also very large which means that they produce a significant amount of waste. That waste (along with uneaten fish food) accumulates in the bottom of your tank where it begins to decompose, producing toxic ammonia. Weekly water changes are essential to remove that accumulated debris (and the toxins) to keep the water quality in your tank high. Clean water means happy fish which means fewer problems with aggression.

Maintaining peace and harmony in a tank full of aggressive fish is by no means an easy task. It will take some time and a lot of trial and error to find an aquascaping arrangement that works for all of your fish. Just be sure to keep a close watch on things as your fish are getting acclimated to the tank and don't be afraid to rearrange things a little if you think of a solution that will work better. And don't forget to keep your tank clean and your water quality high – these things will help to reduce stress and tension in your tank which will also help to reduce aggression.

References

- Fish, animal: britannica.com, Retrieved 11, July 2020
- Popular-types-of-pet-fish: scratchpay.com, Retrieved 24, March 2020
- Mitchell-education-fish-habitats: purefishing.com, Retrieved 09, June 2020
- Fish, migration-animal, science: britannica.com, Retrieved 19, February 2020
- Understanding-aggressive-fish: hartz.com, Retrieved 28, May 2020
- Aquascaping-tips-to-reduce-aggression-in-fish: ratemyfishtank.com, Retrieved 02, January 2020

Chapter 2

Fish Taxonomy

Fishes can be categorized into various classes such as Agnatha, Osteichthyes, Chondrichthyes and Sarcopterygii. Jawless fishes come under class Agnatha; Osteichthyes contains bony fishes, cartilaginous fishes fall under Chondrichthyes and Sarcopterygii contains fleshy finned fishes. This chapter has been carefully written to provide an easy understanding of these classes of fishes.

On the Basis of Habitat

Fish can be classified based on its habitat. As we know, fish comes from the sea, ocean, river, or lake, and each of these habitats play a very important role in its taste and texture. No matter where the fish comes from it is still referred to as seafood.

There could be a species of fish that thrives both in sea as well as river. These are then differentiated by their name such as sea sole or river sole. The taste does differ, as sea fish are considered to be healthier because of the salinity in the water. The flavour of such fish is called 'oceanic flavour'.

Freshwater Fish

Some species of fish tolerate and sometimes migrate back and forth between saline waters and fresh waters. Fish that are predominantly inhabitants of fresh water are described as freshwater fish and include bass, perch, pike, smelt, sturgeon, trout, etc. Freshwater fish are lean because they exercise more—swimming against the current and migrating along the river course. The movement of fish in their breeding season also contributes to their being lean.

Seawater Fish

These are inhabitants of marine water and are adapted to the sea conditions. These fish do not exercise much because the high concentration of salt in the sea water helps them float and they do not have to move against any current. This is the reason why most of these fish are oily and fatty in nature. Oils present in some of them, such as Omega-3, are beneficial for heart patients.

The marine fish vary in size, from very small such as anchovies, to very large such as dogfish and sharks. Some other examples of sea fish are codfish (Atlantic cod. Pacific cod, haddock, pollock, silvery pout, Alaskan pollock), cutlass fish, eels, flat fish (flounders, dabs, turbot, plaice, halibut), herrings (Atlantic herring. Pacific herring, sprats, sardines, pilchards), mackerels, mullet, redfish, salmon, tuna (blue fin, albacore, yellow, big eye, skipjack), etc.

On the Basis of Physical Shape

Fish are also classified as per their shape. There are two shapes of fish—round and flat. As the name suggests, flat fish are flat in appearance while round fish are well rounded and plump. There are many ways in which one can differentiate between flat and round fish.

Table: Difference between flat and round fish.

Flat fish	Round fish
• This fish is flat in shape. • Both its eyes are on one side. • It swims flat to the surface of the water with the belly aligned parallel to the water bed. • Belly side of the fish is while in colour. • It yields four fillets. • Example are place, turbot, flounder, skate, halibut, sole, etc.	• This fish is round in shape. • Eyes are on either side of the fish. • It swims in a horizontal fashion. • Both the side of the fish are of same colour. • It yields two fillets. • Example are salmon, trout, snapper, etc.

On the Basis of Flesh Type

Fish are also classified by its flesh type. The flesh of fish is either oily or white. White fish is also referred to as lean fish. All flat fish and many others, such as snapper, cod, etc., are lean fish or white fish. Fish such as herring, mackerel, salmon, sardines, etc. are classified as oily fish. It is very important for chefs to understand these differences in the structure of fish as cooking methods for oily fish and white fish differ.

Agnatha

Agnathan (superclass Agnatha) is any member of the group of primitive jawless fishes that includes the lampreys (order Petromyzoniformes), hagfishes (order Myxiniformes), and several extinct groups.

Lamprey. Lamprey (*Lampetra*) on rainbow trout.

Hagfishes are minor pests of commercial food fisheries of the North Atlantic, but lampreys, because of their parasitic habit, have been a serious pest of food fisheries in the Great Lakes in North America, where they have reduced the numbers of lake trout and other species. Agnathans are otherwise of little economic importance. The group is of great evolutionary interest, however, because it includes the oldest known craniate fossils and because the living agnathans have many primitive characteristics.

The body of the hagfish is soft-skinned, scaleless, and nearly cylindrical, with a single nostril at the anterior end, overlying the mouth, and a low caudal fin around the tail. In some species the gills open to the surface through separate pores; however, in others the gills open to a common duct, which in turn opens to the surface through a single pore. The eyes are vestigial and covered by skin. All of the approximately 70 known species are restricted to cold, marine bottom waters at depths ranging from 10 metres (about 33 feet) in high latitudes to 1,300 metres (about 4,300 feet) in equatorial oceans. Adults are 40 to 80 cm (about 15 to 30 inches) long. All species are superficially similar except in the number and position of the gill apertures.

Lampreys, which number about 43 species, are found in cool, fresh, and coastal waters of all continents except Africa. All species are rather similar. The body is smooth, scaleless, and eel-shaped, with well-developed dorsal and caudal fins; the mouth is surrounded by a suctorial oral disk bearing horny teeth. The eyes are well developed, and the single nostril is on the top of the head. Lampreys possess seven pairs of external gills. Adults range from 15 to 100 cm (6 to 39 inches) long.

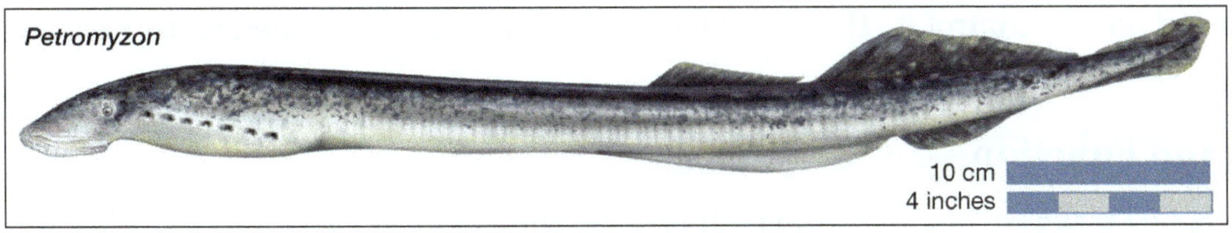

Lamprey: Lamprey (*Petromyzon*).

Although the gonad of a hagfish usually includes both ovary and testis, there is no evidence of either hermaphroditism (the reproductive organs of both sexes functioning in the same individual) or self-fertilization. A female produces a small number of tough-skinned yolk-filled eggs about 2 cm (0.8 inch) long that hatch into miniature adults.

Hagfishes locate their food by scent. Although some are known to eat fishes immobilized in nets, the best-studied species, *Myxine glutinosa*, normally feeds on soft-bodied invertebrates and larger dead animals. *Myxine* burrows into soft marine sediments and rests with only the tip of the head protruding. During respiration, water enters through the nostril and passes by a nasopharyngeal duct to the pharynx and gills. When stimulated by the scent of a dead fish, *Myxine* leaves its burrow and swims against the current. On making contact with the fish, *Myxine* coils around it and bites into it by protruding and retracting the comblike horny tooth plates on the floor of the mouth. Along each side of *Myxine*'s body is a row of prominent glands that produce a gelatinous slime when it is disturbed.

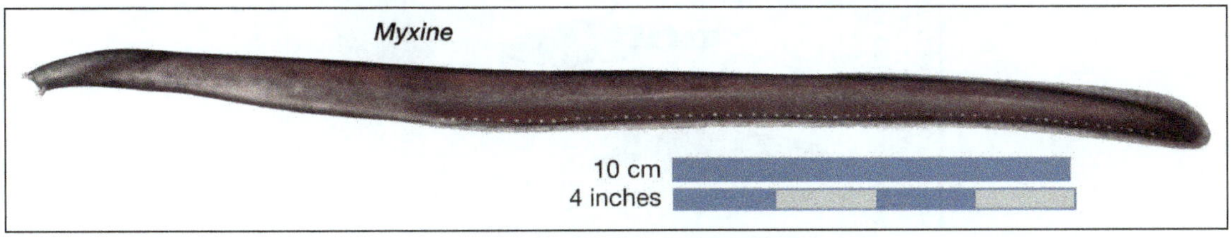

Hagfish: Hagfish (*Myxine*).

Lampreys mate in a nestlike depression excavated by the male in the gravel bed of a stream; the numerous eggs, about 1 mm (0.04 inch) in diameter, lodge in the gravel around the nest. The egg hatches into an ammocoete larva—a blind, wormlike animal that burrows in silt. The larva's mouth is overhung by a hoodlike upper lip that protrudes above the surface of the silt. A continuous stream of water passes in through the mouth and out through the seven pairs of gills. Microscopic plants, the food of the ammocoete, are filtered from this respiratory current by strands of mucus produced by the endostyle, which is a gland in the floor of the pharynx.

After about three years, when the ammocoete has grown to about 10 cm (about 4 inches), it undergoes a radical metamorphosis. The eyes complete their development; the upper lip becomes transformed into a suctorial oral disk; the endostyle changes into a thyroid gland; and the fins along the back increase in height. On completion of metamorphosis, a typical lamprey such as *Petromyzon marinus* migrates to the sea, where it feeds by attaching itself with its sucker to bony fishes. It rasps into the flesh with a toothed, tonguelike structure on the floor of the mouth. Saliva containing an anticoagulant facilitates the ingestion of blood and muscle tissue. On attaining full adult size, the lamprey ceases to feed, migrates upstream to a spawning ground, mates, and dies.

Members of several lamprey species do not migrate to sea but feed in fresh water. *Petromyzon marinus dorsatus* once seriously affected commercial fishing in the Great Lakes until measures were undertaken to control it. The brook lampreys do not feed after metamorphosis but mature sexually, reproduce, and die.

Form and Function

The elongated bodies of hagfishes and lampreys are adaptations to a burrowing habit, throughout life in the hagfish but only during the larval period in the lamprey. Considerable variation in body form occurs among extinct agnathans. Some Osteostraci, for example, possessed a flattened head shield and, although possessing a powerful swimming tail, appear to have been bottom dwellers. The size and shape of the mouth suggest that they were filter feeders. The laterally compressed, fishlike form of the anaspids (such as *Pharyngolepis*) indicates a free-swimming habit. The extinct heterostracans include obvious bottom dwellers (*Drepanaspis*) and others (*Pteraspis*) that were apparently adapted to mid-water, or nektonic, life. Some heterostracans, for example, had movable enamel plates inside the lower lip, probably to provide a biting or grazing mechanism. Osteostracans, anaspids, hagfishes, and lampreys have one median nostril, but heterostracans appear to have had two, one at each corner of the mouth.

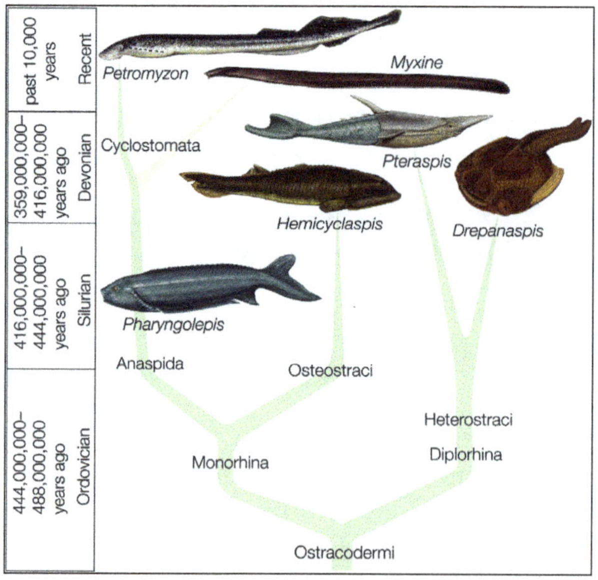

Systematic division of fossil and recent Agnatha.

Evolution and Classification

Evolution

If evidence from fossil and living forms is combined, the Agnatha are distinguishable from the other craniates (Gnathostomata) by what they lack: jaws, lateral fins supported by fin rays, vertebrae, a horizontal semicircular canal in the ear, and genital ducts. The nervous, sensory, endocrine, circulatory, excretory, and muscular systems have the same basic structure as those of gnathostomes but are generally simpler. Unlike the slitlike gills of gnathostomes, the gills of agnathans are pouchlike and open through pores. The presence in the ammocoete larva of an endostyle, a gland that otherwise is found only in protochordates (such as the amphioxi and tunicates), suggests that the Agnatha represent an evolutionary level intermediate between the protochordates and gnathostomes; however, the degree of specialization of known agnathans (particularly the single nostril of most forms) rules out the possibility that they are ancestral to the gnathostomes.

Annotated Classification

In the classification below, the groups are extinct and known only from fossils.

- Superclass Agnatha: Craniate chordates, with pouchlike gills and lacking jaws. Ordovician Period to present.
  - Order Petromyzontiformes (lampreys): 7 pairs of gills opening through pores, laterally placed eyes, single nostril dorsal, horny teeth on an oral sucker, horny teeth on tongue. 1 or 2 dorsal fins. 10 genera, 43 species. Pennsylvanian Subperiod to present.
    - Family Petromyzontidae: Single horny tooth plate above mouth carrying pointed or rounded teeth. 8 genera and approximately 39 species; Eurasia and North America.
    - Family Mordaciidae: Tooth plate above mouth paired, each being tricuspid. 1 genus, 3 species; eastern Australia and western South America.
    - Family Geotriidae: Single tooth plate above mouth carrying 4 teeth. 1 genus, 1 species; southern Australia, New Zealand, South America.
  - Order Myxiniformes (hagfishes): Nostril at tip of snout, eyes vestigial, tongue with 2 rows of horny teeth, mouth surrounded by barbels, slime glands along body, 1–16 pairs of gills, single fin extending around posterior end of body. Pennsylvanian Subperiod to present.
    - Family Myxinidae: Approximately 7 genera, about 70 species.
    - Subfamily Eptatretinae: 5–15 pairs of gills opening separately to outside. 2 genera, about 43 species; Atlantic, Indian and Pacific oceans.
    - Subfamily Myxininae: 5–7 pairs of gills open inside the body into collecting ducts, which in turn open to the outside through a single pore. 4–5 genera, about 26 species; all oceans.

- Order Heterostraci: Head covered with 2 large bony plates, 1 above and 1 below, separated by a variable number of smaller plates. Single gill opening on each side, body covered with large scales, possibly paired nostrils. Bone lacks enclosed bone cells. 6 or 7 minor taxa with questionable affinities. Early Silurian to Late Devonian (about 444–359 million years ago).

- Order Arandaspida: Eyes placed frontally, head covered in 2 large bony plates separated by small plates which each surround separate gill openings. Bone lacks enclosed bone cells. About 4 genera, 4 species. Middle to Late Ordovician (about 472–444 million years ago).

- Order Astraspida: Head covered with small mushroom-shaped plates, gill openings separate. 3 genera, 3 species. Late Ordovician to Early Silurian (461–428 million years ago).

- Order Osteostraci: Head covered in broad bony shield, undersurface of head covered with tiny scales, gill openings on undersurface, eyes dorsal, nostril median and placed on top of head, areas of bony shield covered with tiny tesserae. About 7 families. Early Silurian to Late Devonian (444–359 million years ago).

- Order Galeaspida: Head covered in broad semicircular head shield that is sometimes drawn out to a pointed snout, eyes dorsal, medium nostril very large, gills on the undersurface. 2 major clades (Polybranchiaspidida and Eugaleaspidiformes) and numerous species. Early Silurian to Late Devonian (444–359 million years ago).

- Order Anaspida: Small streamlined fishes, body and head covered with elongate scales, nostril between eyes, gill openings lateral and arranged in slanting line, hypocercal tail bent downward. 3 families. Early Silurian to Late Devonian (444–359 million years ago).

- Order Thelodonti: A little-known group of unknown affinities. Body covered in tiny scales; in some the body is flattened from top to bottom, in others from side to side. About 7 families. Early Silurian to Late Devonian (444–359 million years ago).

Critical Appraisal

Superclass Agnatha is made up of living and extinct forms. In older classifications, extant forms—that is, the lampreys and hagfishes— were grouped together as Cyclostomata (round mouth). It is now thought that the similarities between lampreys and hagfishes are primitive vertebrate features not necessarily indicative of a close relationship. Lampreys share many similarities with jawed vertebrates (gnathostomes)—such as the ability to regulate the ion concentration in the body, the nervous regulation of heartbeat, and well-developed lateral eyes. The extinct agnathans appear to be more advanced than either hagfishes or lampreys and possess features indicative of a relatively close relationship with gnathostomes.

Osteichthyes

Osteichthyes, known as the bony fish, are a taxonomic class (or superclass) of fish and the largest class of vertebrates in existence today. With over 26,000 species, they comprise over 95 percent of all fish species. The Osteichthyes include the ray-finned fish (subclass or class Actinopterygii) and lobe finned fish (subclass or class Sarcopterygii). However, some taxonomic schemes do not consider Osteichthyes a formal taxonomic group, as it is not monophyletic, although the term is retained for vernacular use for those actinopterygians and sarcopterygians conventionally termed fishes.

The members of Osteichthyes represent an extraordinary diversity of forms, including the "living fossil" coelacanths, long "sea-serpent" oarfish, tiny cichlids, whiskered catfish, colorful clownfish, voracious piranhas, unique seahorses, and so forth. This diversity of forms, behaviors, and colors provides aesthetic joy to humans, as well as a variety of practical uses—food, sportfishing, aquarium fish, and so forth. Although lampreys, sharks, and rays are also fish, some individuals commonly use the term only in reference to the jawed bony fish, that is, members of Osteichthyes.

Classification of Osteichthyes

Common Clownfish *(Amphiprion ocellaris)* in their magnificent sea anemone (Heteractis magnifica) home.

Fish are a very diverse assemblage, so much so that the term *fish* is itself more one of convenience than a taxonomic rank. In general, a fish can be defined as a poikilothermic (cold-blooded), water dwelling vertebrate with gills throughout life, and limbs—if present—in the form of fins. Poikilothermic refers to the fact that the internal temperatures of fish vary, often matching the ambient temperature of the environment. (However, some fish, such as the blue-fin tuna, can flourish at significantly higher internal temperatures than the surrounding water.)

Modern fish are conventionally divided into the *jawless fish* (class or superclass Agnatha, with about 75 species including lampreys and hagfish), the *cartilaginous fish* (class Chondrichthyes, with about 800 species including sharks and rays), and the *bony fish* (class or superclass *Osteichthyes*, with over 26,000 species). Some individuals do not include as fish those members of Agnatha or Chondrichthyes.

With over 27,000 known extant (living) species, fish constitute more than one-half of the total vertebrate species. The vast majority of fish are members of Osteichthyes. Nelson recognizes 45 orders, 435 families, 4079 genera, and 23,689 species of osteichthyans (bony fishes).

Within the Osteichthyes, two extant subclasses (or classes) are generally recognized: the Sarcopterygii (lobe-finned fish) and the Actinopterygii (ray-finned or spiny-finned fish). Most bony-fish belong to the Actinopterygii; there are only eight living species of lobe finned fish, including the lungfish (6 extant species) and coelacanths (2 extant species). Some species of lobe-finned fish have jointed bones. Actinopterygii, or ray-finned fish, include the most familiar fish, such as sturgeons, gars, eels, carp, herrings, anchovies, catfishes, goldfishes, piranhas, oarfish, seahorses, bass, cichlids, pickerel, salmon, and trout.

The Actinopterygii are generally classified into two groups—the Chondrostei and the Neopterygii—the latter of which includes the Teleostei (modern bony fishes), a classification into which most fish fit today. The Sarcopterygii are traditionally divided into the lungfish and coelacanths.

While Osteichthyes traditionally is treated as a class of vertebrates, with subclasses Actinopterygii and Sarcopterygii, some newer schemes divide this group into separate classes. Nelson (1994), for example, lists Actinopterygii and Sarcopterygii as classes.

However, while using Osteichthyes as vernacular for actinopterygians and those sarcopterygians conventionally termed fishes, Nelson prefers not to use the term as a formal taxon as it "is clearly not a monophyletic group." Rather, Nelson uses the term "Euteleostomi" for Sacropterygii and Actinopterygii, while Nelson places Actinopterygii and Sarcopterygii (along with the extinct Acanthodii) in grade Teleostomi. Furthermore, Nelson (2006) places the tetrapods (mammals, birds, reptiles, amphibians) in class Sarcopterygii in order to make this a monophyletic group.

Characteristics of Osteichthyes

Osteichthians are characterized by a relatively stable pattern of cranial bones, rooted teeth, and medial insertion of mandibular muscle in lower jaw. The head and pectoral girdles are covered with large dermal bones. The eyeball is supported by a sclerotic ring of four small bones, but this characteristic is missing or modified in many modern species.

The labyrinth in the inner ear contains large otoliths. The braincase, or neurocranium, is frequently divided into anterior and posterior sections divided by fissure. Osteichthyans have a lung or swim bladder. They do not have fin spines, but instead support the fin with lepidotrichia (bone fin rays). They also have an operculum, which helps them breathe without having to swim.

The fin-limbs of sarcopterygiians show such a strong similarity to the expected ancestral form of tetrapod limbs that they have been universally considered the direct ancestors of tetrapods (four-legged vertebrates) in the scientific literature.

Replacement Bone of Osteichthyes

One of the best-known innovations of the osteichthians is endochondral bone or "replacement" bone, i.e. bone ossified internally, by replacement of cartilage, as well as perichondrally, as "spongy bone."

In the more general vertebrates, there are various types of calcified tissue: dentine, enamel (or "enameloids"), and bone, plus variants characterized by their ontogeny, chemistry, form and location. But endochondral bone is unique because it begins life as cartilage.

In more basal vertebrates, cartilaginous structures can become superficially calcified. However, in osteichthians, the circulatory system invades the cartilaginous matrix. This permits the local osteoblasts (bone-forming cells) to continue bone formation within the cartilage and also recruits additional, circulating osteoblasts. Other cells gradually eat away at the surrounding cartilage. The net result is that the cartilage is replaced from within by a somewhat irregular vascularized network of bone. Structurally, the effect is to create a relatively lightweight, flexible, "spongy" bone interior, surrounded by an outline of dense, lamellar periostial bone. Since this bone now surrounds other bone, rather than cartilage, it is referred to as periostial rather than perichondral. This is the unique endochondral bone from which the osteichthians derived their name, as well as many structural advantages.

However useful endochondral bone may be, it is also much heavier and less flexible than cartilage. Thus, many modern osteichian groups, including the extremely successful teleosts, have evolved away from extensive use of endochondral bone.

Examples

The ocean sunfish is the most massive bony fish in the world (but not the longest one; that honor goes to the oarfish). Specimens of ocean sunfish have been observed up to 3.33 m (11 ft) in length and weighing up to 2,300 kg (5,070 lb). Other very large bony fish include the Atlantic blue marlin, some specimens of which have been recorded as in excess of 820 kilograms (1,807.4 lb.), the black marlin, and some sturgeon species.

Chondrichthyes

Chondrichthyan, (class Chondrichthyes), also called chondrichthian is any member of the diverse group of cartilaginous fishes that includes the sharks, skates, rays, and chimaeras. The class is one of the two great groups of living fishes, the other being the osteichthians, or bony fishes. The name Selachii is also sometimes used for the group containing the sharks.

Southern stingray (*Dasyatis americana* or *Hypanus americanus*): Southern stingrays (*Dasyatis americana* or *Hypanus americanus*) are found in the western Atlantic Ocean.

Many unique structural, physiological, biochemical, and behavioral characters make these fishes of particular interest to scientists. The dissection of a small shark is often the biology student's introduction to vertebrate anatomy. These fishes are, in a sense, living fossils, for many of the living sharks and rays are assigned to the same genera as species that swam the Cretaceous seas over 100 million years ago. More than 400 species of sharks and about 500 species of rays are known. Although by any reckoning a successful group, the modern chondrichthyans number far fewer species than the more advanced bony fishes, or teleosts.

The danger some sharks and stingrays present to humans makes these animals fascinating and, at the same time, fearsome. Perhaps for this reason, they figure prominently in the folklore and art of many tropical peoples who depend on the sea. The danger from shark attack, while very real, is remarkably uncommon and easily sensationalized. Quite frequently, little attempt is made to distinguish between dangerous and harmless species.

General Features

Problems of Taxonomy

The name Selachii refers to a category of fishlike vertebrates characterized by a skeleton primarily composed of cartilage. Selachii are given a variety of treatments by ichthyologists. Some authorities consider Selachii to be a class or subclass that contains all the modern sharks and rays; other authorities restrict the use of the name to an order of modern sharks and certain extinct ancestral forms. Under the latter system, the rays (including the sawfishes, guitarfishes, electric rays, mantas, skates, and stingrays) are ranked separately.

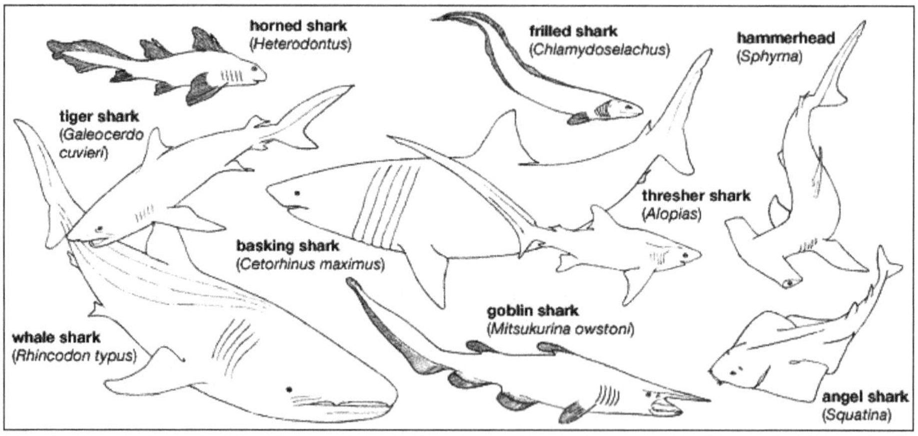

Body plans of representative sharks.

The chimaeras (Holocephali) bear many similarities to sharks and rays in skeletal structure, internal organs, and physiology. Ichthyologists commonly although not unanimously emphasize these similarities by grouping the modern and ancient sharks, rays, and chimaeras in the class Chondrichthyes, the cartilaginous fishes. Under this system, the sharks, skates, and rays are further grouped into one subclass, Elasmobranchii, and the chimaeras into another, Holocephali. Some authorities classify the elasmobranchs into one class (Selachii) and classify the chimaeras into another (Holocephali); however, assigning the two groups class rank implies a degree of distinctness equal to that of the amphibians (Amphibia), reptiles (Reptilia), birds (Aves), and mammals (Mammalia).

Distribution and Abundance

The majority of sharks and rays are marine fishes, but many enter estuaries; some travel far up rivers, and a few are permanent residents of fresh water. Most species live in the relatively shallow waters of continental margins or around offshore islands; a few roam far out in the vast spaces of the oceans. Some live at great depths, in midwaters or on the bottom; others are surface swimmers or inhabit the bottom in shallow waters.

Once regarded as "trash" fish, sharks and rays are increasingly represented in the fisheries of most countries. With numbers of more highly valued bony fishes decreasingly rapidly, many fisheries are specifically targeting elasmobranchs as a primary commercial resource. Annual yields are as much as 750,000 metric tons (roughly 827,000 short tons), and most of this is sold fresh, dried, salted, or processed. This intense harvest is causing the reduction of many shark and ray populations and, in some cases, causing the commercial extinction of some species. One consequence of this depletion of these top-level predators may be a disruption of the food chain in marine ecosystems.

Economic Uses of Elasmobranchs

Sharks as Food

The meat of sharks is marketed for food in all maritime countries. It may be prepared in various ways—fresh, salted, smoked, or pickled—offered in such forms as steaks, fillets, or flakes and under such misleading names as whitefish, grayfish, swordfish, sea bass, and halibut. The flesh is often rather strong tasting; however, this quality is one that can be removed by cleaning and washing and soaking the flesh in brine.

Since ancient times, Chinese people have used the fins of certain sharks and rays as the basis of an epicurean soup. To meet the demand for this product, they have imported fins from far distant countries. The fins are prepared for market by removing the skin and flesh, leaving only the gelatin-rich cartilaginous rays, which are dried before shipment.

Shark liver oil is used in various regions for tanning leather; for preserving wood; as a lubricant; as a folk medicine against rheumatism, burns, and coughs; as a general tonic; as a laxative; and as an ingredient of cosmetics. The liver of a basking shark (Cetorhinus maximus) yields 80 to 600 gallons (about 300 to 2,300 litres) of oil, which was used in lamps until petroleum products replaced animal oils for illumination. The discovery about 1940 that the liver of the soupfin shark (Galeorhinus galeus) of California is peculiarly rich in vitamin A led to the explosive development of a special fishery in California for this species. It also prompted a search in other parts of the world for sharks having livers of comparable potency. Within a few years, however, the economic bubble burst with the invention of a method for manufacturing synthetic vitamin A. The Australian school shark, which was used originally for vitamin A, is now caught for fish fillets.

Other Shark Products

The hard scales provide an abrasive surface to the skin of sharks and some rays, giving it a special value, as leather called shagreen, for polishing hard wood. When heated and polished, shagreen is used for decorating ornaments and, in Japan, for covering sword hilts.

Shark leather is made in several countries, including the United States, from the skin of certain shark species after removal of the scales by a chemical process. A luxury product, much more durable than cowhide, shark leather is used for footwear, belts, wallets, and other accessories. The most suitable skins for leather are from tiger sharks (*Galeocerdo cuvier*), dusky sharks (*Carcharhinus obscurus*), sandbar sharks (*C. plumbeus*), blacktips (*C. brevipina, C. tilstoni, and C. limbatus*), sandtiger sharks (*Carcharias taurus*), and nurse sharks (*Ginglymostoma cirratum, Nebrius ferrugineus, and Pseudognglymostoma brevicaudatum*).

In Greenland, some Inuit make rope from strips of the skin of the sleeper shark (*Somniosus microcephalus*). Polynesians once added to the effectiveness of their war clubs with sharks' teeth. Sharks' teeth have some commercial value as curios. Traditionally the Maori of New Zealand prized the teeth of the mako shark (*Isurus*), which they wore as earrings.

Economic Value of Rays

About 126,000 short tons (roughly 114,000,000 kilograms) of rays are marketed for food in various countries about the world, principally in Europe and Asia. By-products in local demand are skins of scaleless species for drumheads; those of scaly species are used for shagreen. Livers are used for oil, fins for gelatin. People of many tropical regions—Polynesia, Oceania, Malaysia, Central America, and Africa—have used the spines of stingrays for such items as needles and awls, spear tips and daggers, and for the poison they contain. The entire tail of stingrays, complete with spines, has been used as a whip in various tropical areas.

The electric rays, or numbfish, have little commercial value. The ancient Greeks and Romans used the electric shock of *Torpedo* to relieve diseases of the spleen, chronic headaches, and gout. From the Greek word for electric ray, *narke*, comes the word *narcotic*. Today these fishes are worrisome to bathers who step on them and to fishers who may be shocked when hauling in their wet nets.

Danger to Human Life

Among the more than 400 known shark species, about 30 have been authoritatively implicated in unprovoked attacks on persons or boats; of these only about 15 species are considered dangerous, however. Hospital and other records attest to many attacks on bathers, divers, and people awash in the sea following sea or air disasters. There are also many documented cases of sharks attacking small boats. Many surviving victims have been able to identify the attacking animal as a shark; a few even reported the type of shark, such as a hammerhead (*Sphyrna*). In many instances, witnesses have seen the assailant clearly enough to determine the species. Fragments of teeth left in wounds of victims or in the planking of boats have often been large enough to provide ichthyologists with the means for precise identification; furthermore, there are cases where human-body fragments have been found in sharks caught, killed, and autopsied.

In 1958 the American Institute of Biological Sciences established a Shark Research Panel at the Smithsonian Institution and Cornell University to gather historical and current records of shark attacks throughout the world. For the 35 years from 1928 to 1962, inclusive, the panel listed 670 attacks on persons and 102 on boats. More recently, the International Shark Attack File (ISAF) documented over 1,600 unprovoked attacks between 1960 and 2007. Attacks occur most frequently throughout the year in the tropical zone between latitudes 21 °N and 21 °S; from mid-spring to mid-fall they extend as far north and south as the 42° parallels. For this reason, it was formerly believed that the most dangerous sharks lived in waters warmer than 21 °C (70 °F) and that the risk of attack was greatest in the tropics and in the summer months. It is now thought that this circumstance simply results from the fact that more people swim in warm water. It is known, for example, that the most dangerous shark—the great white shark, or man-eater (*Carcharodon carcharias*)—ranges into the cooler waters of both hemispheres. Two other dangerous species—the tiger shark (*Galeocerdo cuvier*) and the bull shark (*Carcharhinus leucas*)—occur primarily in the tropics.

Along the coasts of Australia, New Zealand, and South Africa and in other areas densely populated by sharks, public beaches have lookout towers, bells or sirens, and nets to protect bathers. Since 1937 Australia has used meshing offshore to catch the sharks. Gill nets suspended between buoys and anchors running parallel to the beach and beyond the breaker line have decreased the danger of attack. The nets enmesh sharks from any direction; although they touch neither the surface nor the bottom and are spaced well apart, they provide effective control. South Africa has used a similar protection system and has also conducted experiments with electrical barriers.

The shark species implicated in attacks on persons or boats are mostly large sharks with large cutting teeth. Size, however, is not a dependable criterion; some smaller sharks may bite or nip a bather, inflicting a minor wound. The largest species, the basking shark and the whale shark (*Rhincodon typus*), which grow to 12 and 18 metres (40 and 60 feet) respectively, subsist on minute planktonic organisms and on small schooling fishes. Although either might charge a boat if provoked, only two records of such occurrences have been reported, both in Scotland and both identified with the basking shark. More than 85 percent of all shark species are too small, too unsuitably toothed, or too sluggish or live at depths too great to be potentially dangerous. The most dangerous sharks include, in addition to the white shark, the hammerheads (*Sphyrna*), tiger (*Galeocerdo*), blue (*Prionace*), and requiem sharks (*Carcharhinus*).

Most stingrays live in shallow coastal waters. Some move with the tides to and from beaches, mud flats, or sand flats. Anyone wading in shallow water where they occur runs some risk of stepping on one and provoking an instant response: the ray lashes back its tail, inflicting an agonizingly painful wound that occasionally leads to fatal complications. Rays can be serious pests to shellfisheries, for they are extremely destructive to oyster and clam beds.

Food Habits

Sharks

All sharks are carnivorous and, with a few exceptions, have broad feeding preferences, governed largely by the size and availability of the prey. The recorded food of the tiger shark (*Galeocerdo cuvier*), for example, includes a wide variety of fishes (including other sharks, skates, and stingrays), sea turtles, birds, sea lions, crustaceans, squid, and even carrion such as dead dogs and garbage thrown from ships. Sleeper sharks (*Somniosus*), which occur mainly in polar and subpolar regions, are known to feed on fishes, small whales, squid, crabs, seals, and carrion from whaling stations. Many bottom-dwelling sharks, such as the smooth dogfishes (*Triakis* and *Mustelus*), take crabs, lobsters, and other crustaceans, as well as small fishes.

The three largest sharks, the whale shark (Rhincodon typus), the basking shark (Cetorhinus maximus), and the megamouth shark (Megachasma pelagios), resemble the baleen whales in feeding mode as well as in size. They feed exclusively or chiefly on minute passively drifting organisms (plankton). To remove these from the water and concentrate them, each of these species is equipped with a special straining apparatus analogous to baleen in whales. The basking shark and the megamouth shark have modified gill rakers, the whale shark elaborate spongy tissue supported by the gill arches. The whale shark also eats small, schooling fishes.

The saw sharks (Pristiophoridae) and sawfishes (Pristidae), though unrelated, both share a specialized mode of feeding that depends on the use of their long bladelike snout, or "saw." Equipped with sharp teeth on its sides, the saw is slashed from side to side, impaling, stunning, or cutting the prey fish. Saw sharks and sawfishes, like most other rays, are bottom inhabitants.

Thresher sharks (*Alopias*) feed on open-water schooling fishes, such as mackerel, herring, and bonito, and on squid. The long upper lobe of the tail, which may be half the total length of the shark, is used to herd the fish (sometimes by flailing the water surface) into a concentrated mass convenient for feeding. Thresher sharks have also been observed to stun larger fish with a rapid strike of the tail.

Most sharks and rays do not school. Individuals are normally solitary and usually come together only to exploit food resources or to mate. During these encounters, some species may show specific dominance structures, usually based on size. Some species, however, will travel in large schools segregated by size, a habit that protects smaller individuals from being eaten by larger ones. Still other species form sex-segregated schools where males and females live in slightly different habitats or depths. When potential prey is discovered, sharks circle it, appearing seemingly out of nowhere and frequently approaching from below. Feeding behaviour is stimulated by increasing numbers and rapid swimming, when three or more sharks appear in the presence of food. Activity soon progresses from tight circling to rapid crisscross passes. Biting habits vary with feeding methods and dentition. Sharks with teeth adapted for shearing and sawing are aided in biting by body motions that include rotation of the whole body, twisting movements of the head, and rapid vibrations of the head. As the shark comes into position, the jaws are protruded, erecting and locking the teeth into position. The bite is extremely powerful; a mako shark (*Isurus*), when attacking a swordfish too large to be swallowed whole, may remove the prey's tail with one bite. Under strong feeding stimuli, the sharks' excitement may intensify into what is termed a feeding frenzy, possibly the result of stimulatory overload, in which not only the prey but also injured members of the feeding pack are devoured.

In most cases, sharks locate food by smell, which is well developed in nearly all species. Sharks also possess other important senses that allow them to find food, and the importance of each sense varies between species. Their lateral line system, a series of sensory pores along the side of the body for detecting vibrations, allows sharks to detect vibrations in the water. Their network of ampullae allows them to sense weak electrical signals given off by prey, and their eyes are often acute enough to discriminate the size, shape, and colour of their prey. The sum of these senses working together makes a well-integrated system for finding prey.

The chimaeras have only one external gill opening. In the adult the skin on each side of the head is smooth and lacks scales; the teeth consist of six pairs of grinding plates. The dorsal fin and spine are erectile. Like male sharks and rays, male chimaeras have claspers that serve to transfer sperm to the female, but in addition they have an erectile clasping device, the tantaculum, in front of each pelvic fin; most species have another such organ on top of the head.

Senses

Although sharks are often said to have a low order of intelligence, they, as well as rays and chimaeras, have survived successfully over a long period of geologic time. They are well equipped to locate prey and their own kind; to direct the course of their seasonal migrations; to discriminate specific localities; to respond to variations of temperature; to react to attractive or repelling substances in the water; and perhaps even to feel objects some distance away from them. They can see, hear, smell, taste, feel, and maintain their equilibrium. The roles of the sense organs have been studied in only a few species, principally sharks, and consequently remain imperfectly understood.

The sense of smell is highly developed and probably the principal means of locating prey and guiding the predator toward it. Given a favourable direction of current, sharks can detect incredibly minute concentrations—fractions of a part per million (that is, less than 1×10^{-6} parts)—of certain substances in the water, such as blood.

Although their eyes are structurally and functionally adapted for seeing, it is believed that their visual acuity in discerning the form and colour of an object varies between species. The importance of sight is relative to the habitat and feeding habits of each species. Fast predatory sharks tend to have more acute vision, and in some deep-diving species the eyes are well developed to maximize detection of ambient light.

The hearing apparatus, located in the auditory capsule of the cranium, includes a system of semicircular canals, which are responsible for maintaining equilibrium. Sharks seem to be remarkably sensitive to sounds of low frequency and to possess extraordinary faculty for directional hearing. Whether hearing is more sensitive than smell has not yet been established.

Sensory organs identified as taste buds are located on the floor, sides, and roof of the mouth and on the throat, as well as on the tongue. Experiments on several species of large sharks indicate that they do discriminate food types—preferring tuna, for example, to other fish species. Under some conditions, however, they become less fastidious, going into a feeding frenzy in which they attack anything, including others of their own kind.

Sensory organs located in the skin of all sharks, rays, and chimaeras receive a variety of information—vibrations of low frequencies, temperature, salinity, pressure, and minute electrical stimuli, such as that produced by another fish in the vicinity. These sensory organs are located in the lateral line system, in groups of pores called ampullar organs, found on the head, snout, and around the jaws, which detect electrical impulses.

Salt and Water Balance

Most marine vertebrates maintain lower concentrations of salts and other chemicals in their blood than are found in seawater. As a result, these animals face a continuous problem of water loss to the environment, because of the tendency of water to move through membranes from regions of low salt concentration to regions of higher concentration. The marine cartilaginous fishes differ from almost all of the bony fishes (except the coelacanths and aestivating lungfishes) in being able to reabsorb in the renal (kidney) tubules most of their nitrogenous waste products (urea and trimethylamine oxide) and to accumulate these products in their tissues and blood, an ability termed the urea retention habitus. The concentration within the body thus exceeds that of the surrounding seawater, and water moves into the body with no expenditure of energy. When any of these fishes moves into fresh water, as many do, the urine flow to the outside increases; hence, the concentration of urea in the blood decreases. In the sawfish, for example, the increase of urine output is more than twentyfold; the blood urea concentration decreases to less than one-third the amount observed in marine forms. Purely freshwater elasmobranchs, such as the stingrays of the Orinoco and Amazon drainage systems, seem to lack the urea retention habitus.

Respiration

Sharks breathe chiefly by opening the mouth while expanding the mouth-throat (bucco-pharyngeal) cavity and contracting the gill pouches to close the gill slits. With the mouth closed, they contract the bucco-pharyngeal cavity while dilating the gill pouches, thus drawing the water over the gills where the exchange of oxygen and carbon dioxide takes place. Then, with the mouth still closed, they contract the bucco-pharyngeal cavity and gill pouches, and the gill slits are opened to expel the water.

Most of the rays, on the other hand, take in water chiefly through the spiracles; these then close by contraction at their anterior margins, which bear rudimentary gill filaments and a spiracular valve. Folds of membrane on the roof and floor of the mouth prevent the water from passing down the throat and direct it to the gill openings. Skates, which sometimes hold the lower surface of the head slightly above the bottom, may inhale some water through the mouth; mantas, which have small spiracles and live near the surface, respire chiefly through the mouth. Skates, stingrays, guitarfishes, and angel sharks frequently reverse the direction of flow through the spiracles, apparently to clear them of foreign matter.

Chimaeras take in water chiefly through the nostrils, keeping the mouth closed for the most part. The water reaches the mouth primarily through grooves leading there from the nostrils.

Reproduction and Development

All species of sharks, rays, and chimaeras produce large yolk-rich eggs. These are fertilized internally, for which the males are equipped with two copulatory organs called claspers along the inner edges of the pelvic fins. Each clasper has a groove for guidance of sperm. The few published descriptions of mating sharks and rays are probably characteristic of the entire group. The male grasps one of the female's pectoral fins with his teeth to hold her in position as he inserts a clasper through a cavity (cloaca) and into a tube (oviduct). Males of most species probably use only one clasper at a time. The sperm travel to the anterior end of the oviduct, where they fertilize the eggs. The eggs then move down the oviduct past the shell gland, where they are covered by a shell or capsule.

Some of the sharks, probably all the skates, possibly some of the guitarfishes, and all of the chimaeras are oviparous (egg-laying species). The eggs are enveloped in a horny shell, usually equipped with tendrils for coiling around solid objects or with spikelike projections for anchoring in mud or sand. The egg cases of most species are more or less pillow-shaped; those of the horned sharks (*Heterodontus francisci*) are screw-shaped with a spiral flange. The eggs of chimaeras are elliptic, spindle-shaped, or tadpole-shaped and open to the exterior through pores and slits that permit entrance of water during incubation. An egg of the whale shark found in the Gulf of Mexico measured 30 cm (12 inches) long by about 14 cm (5.5 inches) wide and was 8 cm (3 inches) thick. Protected by the shell and nourished by the abundant yolk, the embryo of an oviparous species develops for 18 to 59 weeks before hatching.

The majority of sharks and rays other than the skates are ovoviviparous (that is, the egg hatches within the mother). In this case, the egg is first coated in the shell gland with a temporary membranous capsule that lasts only during early development. After emerging from its capsule, the embryo remains in the oviduct of the mother, nourished by the yolk sac to which it remains attached. Embryos of some ovoviviparous sharks, notably the porbeagle (*Lamna nasus*), the mako (*Isurus oxyrinchus*), and the sand shark (*Odontaspis taurus*), ingest yolks of other eggs and even other embryos within the oviduct of the mother after the contents of their own yolk sacs are exhausted. In the majority of ovoviviparous sharks and rays, organically rich uterine secretions provide supplemental nourishment, which is absorbed by the yolk sac and in many cases by appendages borne on its stalk. In some genera of rays, vascular filaments producing these secretions extend through the spiracles and into the digestive tract of the embryos.

Several shark species are viviparous—that is, the yolk sac develops folds and projections that interdigitate with corresponding folds of the uterine wall, thus forming a yolk-sac placenta through which nutrient material is passed from the mother.

Growth

Growth of a few shark species has been measured or estimated by the differences in length at the times of tagging and recapturing specimens. Growth is also measured by the statistical analysis of the length in systematically collected samples, by the space between concentric circles on the centra of the vertebrae, and by periodic measurements of specimens kept in aquariums. All studies indicate a slow growth rate. During the 10 years between birth and maturity, male Atlantic spiny dogfish grow an average of 47 cm (19 inches) and females 67 cm (26 inches). The Greenland shark (*Somniosus microcephalus*), which attains 6.5 metres (21 feet) or more (although rarely taken larger than about 4 metres [13 feet]), grows only about 7.5 mm (about 0.3 inch) per year. The annual growth increments of tagged juvenile whitetip reef and Galapagos sharks, both species that become at least 2.5 metres (8 feet) long, were found to be 31 to 54 mm (1 to 2 inches) and 41 mm (about 1.5 inches), respectively. The Australian school shark (*Galeorhinus australis*) grows about 80 mm (3 inches) in its first year and about 30 mm (1 inch) in its 12th year. By its 22nd year, it is estimated to be approaching its maximum length of 1.6 metres (about 5 feet).

The disk of the eastern Pacific round stingray (*Urolophus halleri*) increases in width on the average from 75 mm (3 inches) at birth to 150 mm (6 inches) when mature (that is, at 2.6 years old). In the next five years it grows about 60 mm (about 2.4 inches) more toward its maximum recorded width of 25 cm (10 inches) in males or 31 cm (12.25 inches) in females. The males of European thornback rays (*Raja clavata*) are about 50 cm (20 inches) wide when they reach first maturity, about seven years after birth; females are 60 to 70 cm (24 to 28 inches) at first maturity, nine years after birth.

Evolution and Classification

Evolution

The earliest fossil remains of fishlike vertebrates are too fragmentary to permit tracing the modern fishes precisely to their origins. It is believed that the ancestral forms evolved toward the end of the Ordovician Period (about 455 million years ago) in the upper reaches of streams. During the end of the Silurian and the beginning of the Devonian periods that followed, there appeared an exceedingly diverse group of armour-plated fishes with jawlike structures, paired fins, and bony skeletal tissue. Paleontologists refer to these extinct forms as a distinct class, Placodermi. Between the beginning and end of the Devonian (the latter about 350 million years ago), the placoderms reached their peak in diversity and numbers and almost completely died out; only a few lingered another 10 million years into the Mississippian subperiod (roughly, the Early Carboniferous). During their flowering, the placoderms evidently gave rise to the Osteichthyes (the bony fishes) and the Chondrichthyes (the cartilaginous fishes). Even though the lines of evolution remain to be discovered, it seems quite clear that the two groups evolved independently, the Chondrichthyes appearing much later than the Osteichthyes.

Chondrichthians; *Cladoselache*; *Hybodus:* Two extinct chondrichthians.

Although a few sharklike forms remained in freshwater environments, the vast majority soon invaded the sea, perhaps in response to the arid Devonian climate. There they adapted to life in salt water by evolving the urea retention habitus. Their cartilaginous skeleton, far from representing an evolutionary stage antecedent to the Osteichthyes, as was once believed, is more than likely degenerate rather than primitive. Possibly their precursors were the petalichthyids, a group of Devonian sharklike placoderms that had ossified skeletons and well-developed fins.

The phyletic relationship of the chimaeras and the sharks and rays is a subject open to varying interpretation. Although both groups have many characteristics in common (such as the possession of a cartilaginous skeleton, placoid scales, teeth simply embedded in gums, a spiral valve in the intestine, urea retention habitus, internal fertilization [for which the males have claspers], and the absence of a swim bladder), the two groups may have evolved independently along parallel lines. The chimaeras evolved from the pyctodonts, an order of Devonian placoderms with body form and tooth structure very suggestive of modern chimaeras.

The first fishes clearly identified with the Chondrichthyes were sharklike in form. One order, the Pleurocanthodii, made up of one family of freshwater sharklike fishes, appeared in the Late Devonian (about 380 million years ago). Pleurocanthodians were abundant in the Carboniferous and Early Permian (an interval lasting from 360 million to about 270 million years ago); however, they disappeared during the Triassic Period, which followed. These fishes possessed a skeletal structure made up of both pectoral and pelvic fins that had an axis with side branches (called the archipterygial type). The pleurocanthodian tail was almost symmetrical, being only slightly tilted upward. In addition, a long movable spine projected backward from the back of the head, the teeth had two divergent prongs and a central cusp set on a buttonlike base, the anal fin was two-lobed, and the males had claspers.

The other order, Cladoselachii, consisted of marine fishes known only from fossils of the late Middle Devonian, Carboniferous, and Early Permian periods. In the members of this order, each tooth had a long base composed of a bonelike tissue. From this bonelike tissue, three conical cusps, tall central one and two smaller ones, one on either side, arose. The body scales also had several lobes or cusps; however, the cladoselachians lacked claspers. The jaws had double articulation, extending forward to the snout, and the outline of the caudal (tail) fin was almost symmetrical but with differing internal structure of the upper and the lower lobes.

The cladoselachians were probably ancestral to a group closer to modern sharks, the order Hybodontii. They probably represent an intermediate state in selachian evolution and are classified by some authorities in the order Selachii. Although the jaws had the primitive double articulation, the skeletal support of the pectoral and pelvic fins was close to that of modern selachians, with basal elements projecting outward into the fins. The teeth near the front of the mouth were generally sharp-cusped, and the cusps of those farther back were sometimes reduced to a rounded crown. The front teeth were suitable for seizing prey, whereas those in the back were suitable for crushing mollusks. The hybodonts appeared toward the end of the Devonian, flourished in the Late Paleozoic, and died out during the latter half of the Mesozoic, a few lasting into the Late Cretaceous (about 80 million years ago).

The great period of radiation (diversification) in marine vertebrates characterizing the Paleozoic ended in the Permian. At this time the chondrichthyan fishes, which had reached their greatest flowering during the Carboniferous, became greatly reduced. They remained so until the Jurassic (about 190 million years ago), when the areas of the seas expanded and those of the land diminished. The six-gilled shark (*Hexanchus*), horned shark (*Heterodontus*), and guitarfishes appeared during the Jurassic. By the end of the Cretaceous (about 65 million years ago), most of the families and many genera of modern sharks, skates, and rays were represented. The evolution of elasmobranch fishes, much as they are known today, had been accomplished.

Sarcopterygii

Sarcopterygii traditionally is a class of vertebrates known as lobe-finned fishes, consisting of living and fossil lungfishes and coelacanths, and related extinct fishes. Members of this group are characterized by lobed paired fins, joined to the body by a single bone and two dorsal fins with separate bases.

Among the coelacanths, the first living species, Latimeria chalumnae, was found in 1938, although diverse fossil specimens (now placed in several families) were well known at the time. Two living species are currently known. The dipnoans (lungfishes) also are well represented in the fossil record, but there are only six extant species, with the first living lungfish formally described in 1837.

In order to make the taxon monophyletic, the tetrapods are sometimes included as part of class Sarcopterygii; that is, fishes in this group are considered more closely related to mammals and other tetrapods than to other fishes. The fact that lungfish would be placed in the same class as such diverse animals as eagles, alligators, and chimpanzees, but not with sharks or salmon, shows the emphasis placed on lineage in current taxonomy.

Sarcopterygii is traditionally viewed as a class of fishes, including the lungfishes and the coelacanths. Historically, it has been considered, along with Actinopterygii, one of two extant (living) classes of fishes in the taxon Osteichthyes, or bony fishes, characterized by their bony skeleton instead of cartilage. However, Nelson, in his authoritative *Fishes of the World* discontinues Osteichthyes for formal taxonomic use and instead places Actinopterygii and Sarcopterygii (along with the extinct Acanthodii) in grade Teleostomi. The term osteichthyians is retained for vernacular use for those actinopterygians and sarcopterygians conventionally termed fishes.

Some taxonomists, who continue the use of the term Osteichthyes, nonetheless consider Sarcopterygii a sister superclass to Osteichthyes, instead of a class below them, based on the vast differences between Sarcopterygii and Osteichthyes in fin structure, respiratory structure, and circulatory structure.

In order to make Sacropterygii a monophyletic group, includes as part of Sarcopterygii not only the fishes, but also all the tetrapods (amphibians, reptiles, birds, and mammals). This is the approach of many taxonomists who subscribe to the cladistic approach. Indeed, the fin limbs of sarcopterygiians show such a strong similarity to the expected ancestral form of tetrapod limbs that they typically are considered the direct ancestors of tetrapods in the scientific literature. Thus, while there are eight extant species of fishes in Sarcopterygii, Nelson recognizes almost 27,000 extant species in this class. Traditionally, the term Sarcopterygii was used to include only fishes called dipnoans (lobe-finned fishes) and crossopterygians.

There are various taxonomic schemes for the class Sarcopterygii. recognizes two subclasses of Sarcopterygii: Coelacanthimorpha and Dipnotetrapodomorpha. Coelacanthimorpha includes one order, Coelacanthiformes (coelacanths), with eight families that include only fossil members and one family that includes two living species of coelacanth as well as fossil members. The subclass Dipnotetrapodomopha is a new term that includes several orders and families, with the order Ceratodontiformes including the three extant families, three extant genera, and six modern species.

Characteristics of Osteichthyes

Sarcopterygians are bony fish with lobed paired fins, which are joined to the body by a single bone. These fins are considered to have evolved into the legs of the first tetrapod land vertebrates, amphibians. Sarcopterygians also possess two dorsal fins with separate bases, as opposed to the single dorsal fin of actinopterygians (ray-finned fishes). The braincase of sarcoptergygians primitively has a hinge line, but this is lost in tetrapods and lungfish. Many early sarcopts have a symmetrical tail.

Coelacanths (order Coelacanthiformes), living and extinct, are characterized by a caudal fin with three lobes, external nostrils, and an anterior dorsal fin in front of the center of the body. There are two living species, *Latimeria chalumne* and *L. menadoensis*. The first living species ever found, *Latimeria chalumnae*, is a marine species known from South Africa, Comoros Archipelago, and the Mozambique area. It was first found off of South Africa in 1938. Adults reach 1.8 meters and it has an unusual method of swimming in keeping its body rigid. It is the only living chordate with an intracranial joint, although this feature appears in other coelacanths.

The lungfishes (dipnoi) have platelike teeth useful for crushing and grinding. The six extant species of lungfishes are placed by Nelson in the order Ceratodontiformes of the superorder Ceratodontimorpha. These are characterized by functional lungs, a caudal fin that confluent with the dorsal and anal fins, and the absence of premaxilla and maxilla. The extant species are all freshwater forms. There is one species of Australian lungfishes (family Ceratodontidae), one species of South American lungfishes (family Lepidosirenidae; found in Brazil and Paraguay), and four species of African lungfishes (family Protopteridae; all placed in the genus *Protopterus*). The air bladders (lung) of the Australian lungfishes are unpaired, but the air bladders of the other families of lungfishes are paired. Likewise, the Australian lungfishes have flipper-like pectoral and pelvic fins, large scales, and larvae without external gills, while the other species have filamentous pectoral and pelvic fins, without rays, small scales, and larvae with external gills.

Evolution of Sarcopterygii

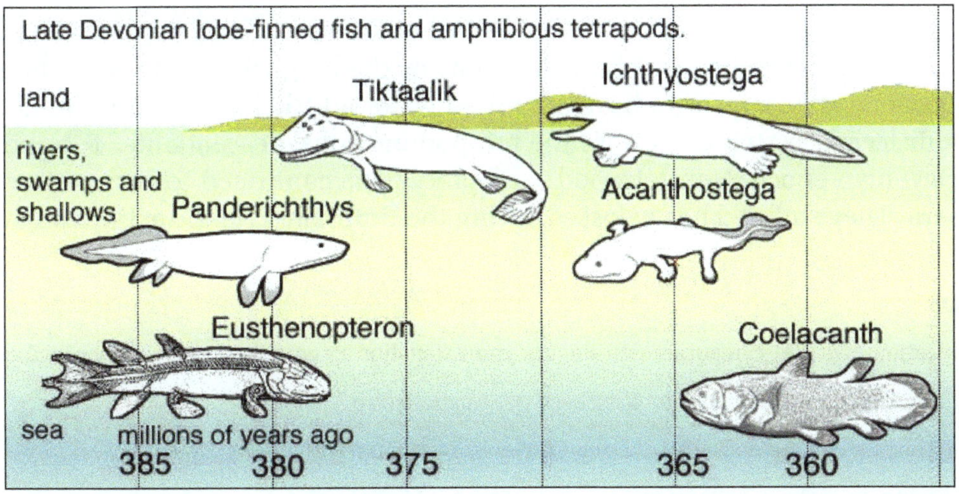

Figure shows in Late Devonian vertebrate speciation, descendants of pelagic lobe-finned fish, like *Eusthenopteron*, exhibited a sequence of adaptations:

- *Panderichthys*, suited to muddy shallows.
- *Tiktaalik* with limb-like fins that could take it onto land.
- Early tetrapods in weed-filled swamps, such as:
  - *Acanthostega*, which had feet with eight digits.
  - *Ichthyostega* with limbs.

Descendants also included pelagic lobe-finned fish such as coelacanth species. The oldest sarcopterygians were found in the Uppermost Silurian. The first Sarcopterygian closely resembled Acanthodians. The sarcopterygians closest relatives at the time were the actinopterygians (ray-finned fishes). Sarcopterygians probably evolved in the oceans, but they later came into freshwater habitats, possibly as an adaptation to avoid the predatory placoderms, which were dominant in the Early–Middle Devonian seas.

During the Early Devonian, the sarcopterygian line split into two main lineages: the coelacanths and the Rhipidistia. The coelacanths appeared in the Early Devonian, and stayed in the oceans. The coelacanths' heyday was the Late Devonian and Carboniferous, as they were more common during those periods than in any other period in the Phanerozoic. Coelacanths still live today in the oceans.

Rhipidistians appeared about the same time as the coelacanths, but unlike them, rhipidistians left the ocean world and migrated into the freshwater habitats; their ancestors probably lived in the oceans near the river mouths (estuaries). The rhipidistians, in turn, split into two major groups: the lungfishes, and the tetrapodomorphs.

The lungfishes' greatest diversity was in the Triassic period, but today, there are fewer than a dozen genera left. The lungfishes evolved the first proto-lungs and proto-limbs. The lungfishes, ancient and modern, are adapted to use their stubby fins (proto-limbs) to walk on land and find new water if their waterhole is depleted, and use their lungs to breathe air and get sufficient oxygen.

The tetrapodomorphs have the same basic anatomy as the lungfishes, who were their closest kin, but the tetrapodomorphs appear to have inhabited water a little longer, until the Late Devonian. Tetrapods, four legged vertebrates, were the terapodomorphs' descendants. Tetrapods appeared in the Late Devonian epoch. Non-tetrapod sarcopterygians continued to survive towards the end of Paleozoic era. They suffered heavy losses during the Permian-Triassic extinction event.

References

- Classification-of-fish-3-categories, fish-86520: yourarticlelibrary.com, Retrieved 26, April 2020
- Agnathan, animal: britannica.com, Retrieved 07, August 2020
- Osteichthyes: newworldencyclopedia.org, Retrieved 16, January 2020
- Chondrichthian, animal: britannica.com, Retrieved 29, May 2020
- Sarcopterygii: newworldencyclopedia.org, Retrieved 17, July 2020

Chapter 3

Anatomical Physiology of Fishes

Fish physiology deals with the study of the component of fishes as well as the way in which they function together. Some of the topics studied under anatomical physiology of fishes are fish jaw, fish gill, fish scale, fish fin, fish locomotion, sensory systems in fish, pain in fish, etc. The chapter closely examines these key aspects of anatomical physiology of fishes to provide an extensive understanding of the subject.

Fish Physiology

When threatened, the toxic pufferfish fills its extremely elastic stomach with water.

Fish physiology is the scientific study of how the component parts of fish function together in the living fish. It can be contrasted with fish anatomy, which is the study of the form or morphology of fishes. In practice, fish anatomy and physiology complement each other, the former dealing with the structure of a fish, its organs or component parts and how they are put together, such as might be observed on the dissecting table or under the microscope, and the later dealing with how those components function together in the living fish.

Respiration

Most fish exchange gases using gills on either side of the pharynx (throat). Gills are tissues which consist of threadlike structures called filaments. These filaments have many functions and "are involved in ion and water transfer as well as oxygen, carbon dioxide, acid and ammonia exchange. Each filament contains a capillary network that provides a large surface area for exchanging oxygen and carbon dioxide. Fish exchange gases by pulling oxygen-rich water through their mouths and pumping it over their gills. In some fish, capillary blood flows in the opposite direction to the water, causing countercurrent exchange. The gills push the oxygen-poor water out through openings in the sides of the pharynx.

Fish from multiple groups can live out of the water for extended time periods. Amphibious fish such as the mudskipper can live and move about on land for up to several days, or live in stagnant or otherwise oxygen depleted water. Many such fish can breathe air via a variety of mechanisms. The skin of anguillid eels may absorb oxygen directly. The buccal cavity of the electric eel may breathe air. Catfish of the families Loricariidae, Callichthyidae, and Scoloplacidae absorb air through their digestive tracts. Lungfish, with the exception of the Australian lungfish, and bichirs have paired lungs similar to those of tetrapods and must surface to gulp fresh air through the mouth and pass spent air out through the gills. Gar and bowfin have a vascularized swim bladder that functions in the same way. Loaches, trahiras, and many catfish breathe by passing air through the gut. Mudskippers breathe by absorbing oxygen across the skin (similar to frogs). A number of fish have evolved so-called accessory breathing organs that extract oxygen from the air. Labyrinth fish (such as gouramis and bettas) have a labyrinth organ above the gills that performs this function. A few other fish have structures resembling labyrinth organs in form and function, most notably snakeheads, pikeheads, and the Clariidae catfish family.

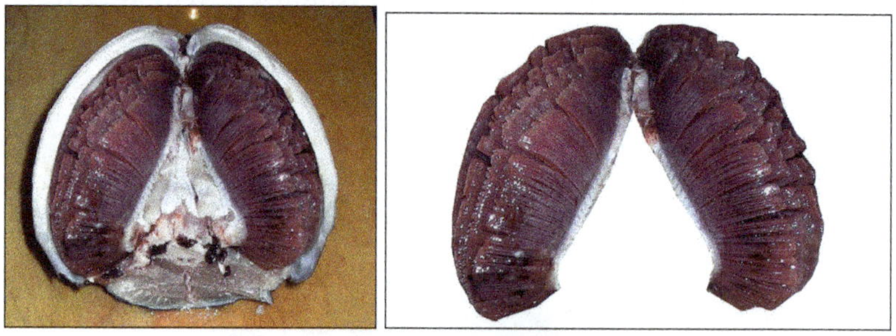

Gills inside the head of a tuna. The head is snout-down, with the view looking towards the mouth. On the right are the detached gills.

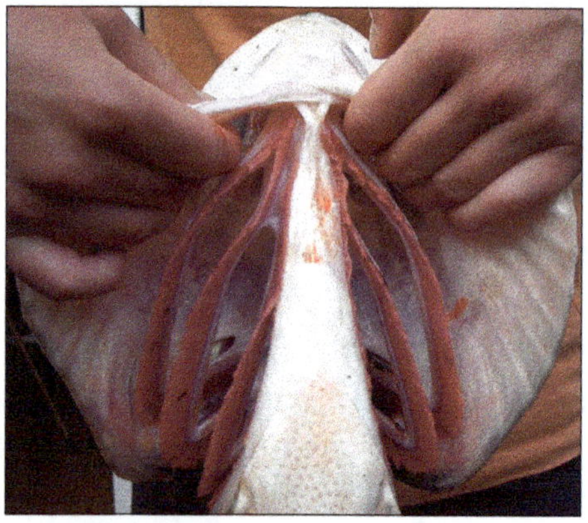

Gill arches bearing gills in a pike.

Breathing air is primarily of use to fish that inhabit shallow, seasonally variable waters where the water's oxygen concentration may seasonally decline. Fish dependent solely on dissolved oxygen, such as perch and cichlids, quickly suffocate, while air-breathers survive for much longer, in some cases in water that is little more than wet mud. At the most extreme, some air-breathing fish are able to survive in damp burrows for weeks without water, entering a state of aestivation (summertime hibernation) until water returns.

Air breathing fish can be divided into *obligate* air breathers and *facultative* air breathers. Obligate air breathers, such as the African lungfish, are obligated to breathe air periodically or they suffocate. Facultative air breathers, such as the catfish *Hypostomus plecostomus*, only breathe air if they need to and can otherwise rely on their gills for oxygen. Most air breathing fish are facultative air breathers that avoid the energetic cost of rising to the surface and the fitness cost of exposure to surface predators.

All basal vertebrates breathe with gills. The gills are carried right behind the head, bordering the posterior margins of a series of openings from the esophagus to the exterior. Each gill is supported by a cartilagenous or bony gill arch. The gills of vertebrates typically develop in the walls of the pharynx, along a series of gill slits opening to the exterior. Most species employ a countercurrent exchange system to enhance the diffusion of substances in and out of the gill, with blood and water flowing in opposite directions to each other.

The gills are composed of comb-like filaments, the gill lamellae, which help increase their surface area for oxygen exchange. When a fish breathes, it draws in a mouthful of water at regular intervals. Then it draws the sides of its throat together, forcing the water through the gill openings, so that it passes over the gills to the outside. The bony fish have three pairs of arches, cartilaginous fish have five to seven pairs, while the primitive jawless fish have seven. The vertebrate ancestor no doubt had more arches, as some of their chordate relatives have more than 50 pairs of gills.

Higher vertebrates do not develop gills, the gill arches form during fetal development, and lay the basis of essential structures such as jaws, the thyroid gland, the larynx, the *columella* (corresponding to the stapes in mammals) and in mammals the malleus and incus. Fish gill slits may be the evolutionary ancestors of the tonsils, thymus gland, and Eustachian tubes, as well as many other structures derived from the embryonic branchial pouches.

Scientists have investigated what part of the body is responsible for maintaining the respiratory rhythm. They found that neurons located in the brainstem of fish are responsible for the genesis of the respiratory rhythm. The position of these neurons is slightly different from the centers of respiratory genesis in mammals but they are located in the same brain compartment, which has caused debates about the homology of respiratory centers between aquatic and terrestrial species. In both aquatic and terrestrial respiration, the exact mechanisms by which neurons can generate this involuntary rhythm are still not completely understood.

Another important feature of the respiratory rhythm is that it is modulated to adapt to the oxygen consumption of the body. As observed in mammals, fish "breathe" faster and heavier when they do physical exercise. The mechanisms by which these changes occur have been strongly debated over more than 100 years between scientists. The authors can be classified in 2 schools:

1. Those who think that the major part of the respiratory changes are pre-programmed in the brain, which would imply that neurons from locomotion centers of the brain connect to respiratory centers in anticipation of movements.
2. Those who think that the major part of the respiratory changes result from the detection of muscle contraction, and that respiration is adapted as a consequence of muscular contraction and oxygen consumption. This would imply that the brain possesses some kind of detection mechanisms that would trigger a respiratory response when muscular contraction occurs.

Many now agree that both mechanisms are probably present and complementary, or working alongside a mechanism that can detect changes in oxygen and/or carbon dioxide blood saturation.

Bony Fish

In bony fish, the gills lie in a branchial chamber covered by a bony operculum. The great majority of bony fish species have five pairs of gills, although a few have lost some over the course of evolution. The operculum can be important in adjusting the pressure of water inside of the pharynx to allow proper ventilation of the gills, so that bony fish do not have to rely on ram ventilation (and hence near constant motion) to breathe. Valves inside the mouth keep the water from escaping.

The gill arches of bony fish typically have no septum, so that the gills alone project from the arch, supported by individual gill rays. Some species retain gill rakers. Though all but the most primitive bony fish lack a spiracle, the pseudobranch associated with it often remains, being located at the base of the operculum. This is, however, often greatly reduced, consisting of a small mass of cells without any remaining gill-like structure.

Marine teleosts also use gills to excrete electrolytes. The gills' large surface area tends to create a problem for fish that seek to regulate the osmolarity of their internal fluids. Saltwater is less dilute than these internal fluids, so saltwater fish lose large quantities of water osmotically through their gills. To regain the water, they drink large amounts of seawater and excrete the salt. Freshwater is more dilute than the internal fluids of fish, however, so freshwater fish gain water osmotically through their gills.

In some primitive bony fishes and amphibians, the larvae bear external gills, branching off from the gill arches. These are reduced in adulthood, their function taken over by the gills proper in fishes and by lungs in most amphibians. Some amphibians retain the external larval gills in adulthood, the complex internal gill system as seen in fish apparently being irrevocably lost very early in the evolution of tetrapods.

Cartilaginous Fish

Like other fish, sharks extract oxygen from seawater as it passes over their gills. Unlike other fish, shark gill slits are not covered, but lie in a row behind the head. A modified slit called a spiracle lies just behind the eye, which assists the shark with taking in water during respiration and plays a major role in bottom–dwelling sharks. Spiracles are reduced or missing in active pelagic sharks. While the shark is moving, water passes through the mouth and over the gills in a process known as "ram ventilation". While at rest, most sharks pump water over their gills to ensure a constant supply of oxygenated water. A small number of species have lost the ability to pump water through their gills and must swim without rest. These species are *obligate ram ventilators* and would presumably asphyxiate if unable to move. Obligate ram ventilation is also true of some pelagic bony fish species.

The respiration and circulation process begins when deoxygenated blood travels to the shark's two-chambered heart. Here the shark pumps blood to its gills via the ventral aorta artery where it branches into afferent brachial arteries. Reoxygenation takes place in the gills and the reoxygenated blood flows into the efferent brachial arteries, which come together to form the dorsal aorta. The blood flows from the dorsal aorta throughout the body. The deoxygenated blood from the body then flows through the posterior cardinal veins and enters the posterior cardinal sinuses. From there blood enters the heart ventricle and the cycle repeats.

Sharks and rays typically have five pairs of gill slits that open directly to the outside of the body, though some more primitive sharks have six or seven pairs. Adjacent slits are separated by a cartilaginous gill arch from which projects a long sheet-like septum, partly supported by a further piece of cartilage called the *gill ray*. The individual lamellae of the gills lie on either side of the septum. The base of the arch may also support gill rakers, small projecting elements that help to filter food from the water.

A smaller opening, the spiracle, lies in the back of the first gill slit. This bears a small *pseudobranch* that resembles a gill in structure, but only receives blood already oxygenated by the true gills. The spiracle is thought to be homologous to the ear opening in higher vertebrates.

Most sharks rely on ram ventilation, forcing water into the mouth and over the gills by rapidly swimming forward. In slow-moving or bottom dwelling species, especially among skates and rays, the spiracle may be enlarged, and the fish breathes by sucking water through this opening, instead of through the mouth.

Chimaeras differ from other cartilagenous fish, having lost both the spiracle and the fifth gill slit. The remaining slits are covered by an operculum, developed from the septum of the gill arch in front of the first gill.

Lampreys and Hagfish

Lampreys and hagfish do not have gill slits as such. Instead, the gills are contained in spherical pouches, with a circular opening to the outside. Like the gill slits of higher fish, each pouch contains two gills. In some cases, the openings may be fused together, effectively forming an operculum. Lampreys have seven pairs of pouches, while hagfishes may have six to fourteen, depending on the species. In the hagfish, the pouches connect with the pharynx internally. In adult lampreys, a separate respiratory tube develops beneath the pharynx proper, separating food and water from respiration by closing a valve at its anterior end.

Circulation

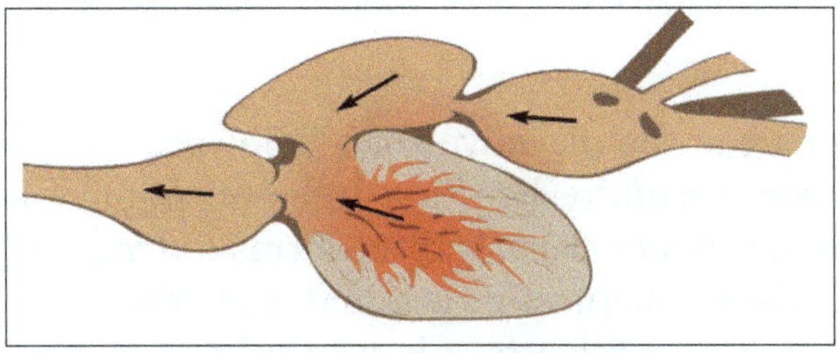

Two-chambered heart of a fish.

The circulatory systems of all vertebrates are *closed*, just as in humans. Still, the systems of fish, amphibians, reptiles, and birds show various stages of the evolution of the circulatory system. In fish, the system has only one circuit, with the blood being pumped through the capillaries of the gills and on to the capillaries of the body tissues. This is known as *single cycle* circulation. The heart of fish is therefore only a single pump (consisting of two chambers). Fish have a closed-loop circulatory system. The heart pumps the blood in a single loop throughout the body. In most fish, the heart consists of four parts, including two chambers and an entrance and exit. The first part is the sinus venosus, a thin-walled sac that collects blood from the fish's veins before allowing it to flow to the second part, the atrium, which is a large muscular chamber. The atrium serves as a one-way antechamber, sends blood to the third part, ventricle. The ventricle is another thick-walled, muscular chamber and it pumps the blood, first to the fourth part, bulbus arteriosus, a large tube, and then out of the heart. The bulbus arteriosus connects to the aorta, through which blood flows to the gills for oxygenation.

In amphibians and most reptiles, a double circulatory system is used, but the heart is not always completely separated into two pumps. Amphibians have a three-chambered heart.

Digestion

Jaws allow fish to eat a wide variety of food, including plants and other organisms. Fish ingest food through the mouth and break it down in the esophagus. In the stomach, food is further digested and, in many fish, processed in finger-shaped pouches called pyloric caeca, which secrete digestive enzymes and absorb nutrients. Organs such as the liver and pancreas add enzymes and various chemicals as the food moves through the digestive tract. The intestine completes the process of digestion and nutrient absorption.

In most vertebrates, digestion is a four-stage process involving the main structures of the digestive tract, starting with ingestion, placing food into the mouth, and concluding with the excretion of undigested material through the anus. From the mouth, the food moves to the stomach, where as bolus it is broken down chemically. It then moves to the intestine, where the process of breaking the food down into simple molecules continues and the results are absorbed as nutrients into the circulatory and lymphatic system.

Although the precise shape and size of the stomach varies widely among different vertebrates, the relative positions of the oesophageal and duodenal openings remain relatively constant. As a result, the organ always curves somewhat to the left before curving back to meet the pyloric sphincter. However, lampreys, hagfishes, chimaeras, lungfishes, and some teleost fish have no stomach at all, with the oesophagus opening directly into the intestine. These animals all consume diets that either require little storage of food, or no pre-digestion with gastric juices, or both.

The small intestine is the part of the digestive tract following the stomach and followed by the large intestine, and is where much of the digestion and absorption of food takes place. In fish, the divisions of the small intestine are not clear, and the terms *anterior* or *proximal* intestine may be used instead of duodenum. The small intestine is found in all teleosts, although its form and length vary enormously between species. In teleosts, it is relatively short, typically around one and a half times the length of the fish's body. It commonly has a number of *pyloric caeca*, small pouch-like structures along its length that help to increase the overall surface area of the organ for digesting food. There is no ileocaecal valve in teleosts, with the boundary between the small intestine and the rectum being marked only by the end of the digestive epithelium.

There is no small intestine as such in non-teleost fish, such as sharks, sturgeons, and lungfish. Instead, the digestive part of the gut forms a spiral intestine, connecting the stomach to the rectum. In this type of gut, the intestine itself is relatively straight, but has a long fold running along the inner surface in a spiral fashion, sometimes for dozens of turns. This valve greatly increases both the surface area and the effective length of the intestine. The lining of the spiral intestine is similar to that of the small intestine in teleosts and non-mammalian tetrapods. In lampreys, the spiral valve is extremely small, possibly because their diet requires little digestion. Hagfish have no spiral valve at all, with digestion occurring for almost the entire length of the intestine, which is not subdivided into different regions.

The large intestine is the last part of the digestive system normally found in vertebrate animals. Its function is to absorb water from the remaining indigestible food matter, and then to pass useless waste material from the body. In fish, there is no true large intestine, but simply a short rectum connecting the end of the digestive part of the gut to the cloaca. In sharks, this includes a *rectal gland* that secretes salt to help the animal maintain osmotic balance with the seawater. The gland somewhat resembles a caecum in structure, but is not a homologous structure.

As with many aquatic animals, most fish release their nitrogenous wastes as ammonia. Some of the wastes diffuse through the gills. Blood wastes are filtered by the kidneys.

Saltwater fish tend to lose water because of osmosis. Their kidneys return water to the body. The reverse happens in freshwater fish: they tend to gain water osmotically. Their kidneys produce dilute urine for excretion. Some fish have specially adapted kidneys that vary in function, allowing them to move from freshwater to saltwater.

In sharks, digestion can take a long time. The food moves from the mouth to a J-shaped stomach, where it is stored and initial digestion occurs. Unwanted items may never get past the stomach, and instead the shark either vomits or turns its stomachs inside out and ejects unwanted items from its mouth. One of the biggest differences between the digestive systems of sharks and mammals is that sharks have much shorter intestines. This short length is achieved by the spiral valve with multiple turns within a single short section instead of a long tube-like intestine. The valve provides a long surface area, requiring food to circulate inside the short gut until fully digested, when remaining waste products pass into the cloaca.

Endocrine system

Regulation of Social Behaviour

Oxytocin is a group of neuropeptides found in most vertebrate. One form of oxytocin functions as a hormone which is associated with human love. In 2012, researchers injected cichlids from the social species *Neolamprologus pulcher*, either with this form of isotocin or with a control saline solution. They found isotocin increased "responsiveness to social information", which suggests "it is a key regulator of social behavior that has evolved and endured since ancient times".

Effects of Pollution

Fish can bioaccumulate pollutants that are discharged into waterways. Estrogenic compounds found in pesticides, birth control, plastics, plants, fungi, bacteria, and synthetic drugs leeched into rivers are affecting the endocrine systems of native species. In Boulder, Colorado, white sucker fish found downstream of a municipal waste water treatment plant exhibit impaired or abnormal sexual development. The fish have been exposed to higher levels of estrogen, and leading to feminized fish. Males display female reproductive organs, and both sexes have reduced fertility, and a higher hatch mortality. The same feminization effect can also be seen in male African clawed frogs when exposed to high levels of atrazine, a widely used pesticide. In the marine ecosystem, organochlorine contaminants like pesticides, herbicides (DDT), and chlordan are accumulating within fish tissue and disrupting their endocrine system. High frequencies of infertility and high levels of organochlorines have been found in bonnethead sharks along the Gulf Coast of Florida. These endocrine-disrupting compounds are similar in structure to naturally occurring hormones in fish. They can modulate hormonal interactions in fish by:

- Binding to cellular receptors, causing unpredictable and abnormal cell activity.
- Blocking receptor sites, inhibiting activity.
- Promoting the creation of extra receptor sites, amplifying the effects of the hormone or compound.
- Interacting with naturally occurring hormones, changing their shape and impact.
- Affecting hormone synthesis or metabolism, causing an improper balance or quantity of hormones.

Osmoregulation

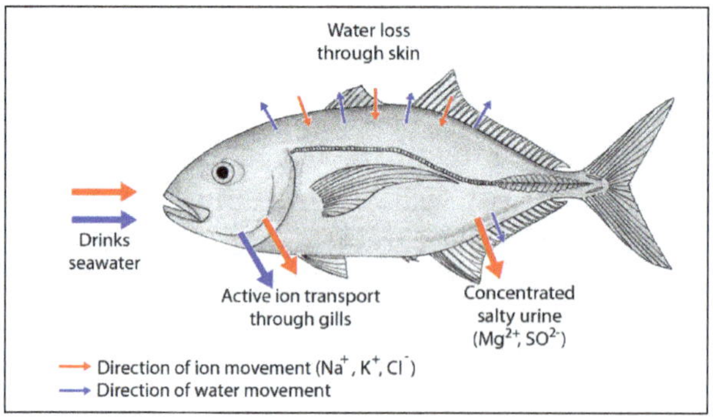

Movement of water and ions in saltwater fish.

Two major types of osmoregulation are osmoconformers and osmoregulators. Osmoconformers match their body osmolarity to their environment actively or passively. Most marine invertebrates are osmoconformers, although their ionic composition may be different from that of seawater.

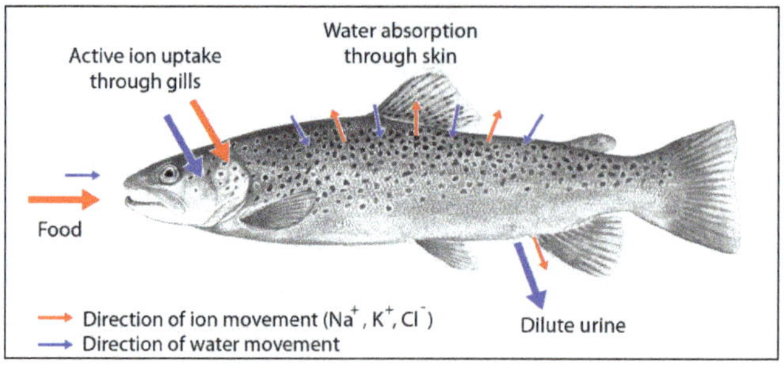

Movement of water and ions in freshwater fish.

Osmoregulators tightly regulate their body osmolarity, which always stays constant, and are more common in the animal kingdom. Osmoregulators actively control salt concentrations despite the salt concentrations in the environment. An example is freshwater fish. The gills actively uptake salt from the environment by the use of mitochondria-rich cells. Water will diffuse into the fish, so it excretes a very hypotonic (dilute) urine to expel all the excess water. A marine fish has an internal osmotic concentration lower than that of the surrounding seawater, so it tends to lose water and gain salt. It actively excretes salt out from the gills. Most fish are stenohaline, which means they are restricted to either salt or fresh water and cannot survive in water with a different salt concentration than they are adapted to. However, some fish show a tremendous ability to effectively osmoregulate across a broad range of salinities; fish with this ability are known as euryhaline species, e.g., salmon. Salmon has been observed to inhabit two utterly disparate environments — marine and fresh water — and it is inherent to adapt to both by bringing in behavioral and physiological modifications.

In contrast to bony fish, with the exception of the coelacanth, the blood and other tissue of sharks and Chondrichthyes is generally isotonic to their marine environments because of the high concentration of urea and trimethylamine N-oxide (TMAO), allowing them to be in osmotic balance with the seawater. This adaptation prevents most sharks from surviving in freshwater, and they are therefore confined to marine environments. A few exceptions exist, such as the bull shark, which has developed a way to change its kidney function to excrete large amounts of urea. When a shark dies, the urea is broken down to ammonia by bacteria, causing the dead body to gradually smell strongly of ammonia.

Sharks have adopted a different, efficient mechanism to conserve water, i.e., osmoregulation. They retain urea in their blood in relatively higher concentration. Urea is damaging to living tissue so, to cope with this problem, some fish retain *trimethylamine oxide*. This provides a better solution to urea's toxicity. Sharks, having slightly higher solute concentration (i.e., above 1000 mOsm which is sea solute concentration), do not drink water like fresh water fish.

Thermoregulation

Homeothermy and poikilothermy refer to how stable an organism's temperature is. Most endothermic organisms are homeothermic, like mammals. However, animals with facultative endothermy are often poikilothermic, meaning their temperature can vary considerably. Similarly, most fish are ectotherms, as all of their heat comes from the surrounding water. However, most are homeotherms because their temperature is very stable.

Most organisms have a preferred temperature range, however some can be acclimated to temperatures colder or warmer than what they are typically used to. An organism's preferred temperature is typically the temperature at which the organism's physiological processes can act at optimal rates. When fish become acclimated to other temperatures, the efficiency of their physiological processes may decrease but will continue to function. This is called the thermal neutral zone at which an organism can survive indefinitely.

H. M. Vernon has done work on the death temperature and paralysis temperature (temperature of heat rigor) of various animals. He found that species of the same class showed very similar temperature values, those from the Amphibia examined being 38.5°C, fish 39°C, Reptilia 45°C, and various Molluscs 46°C.

To cope with low temperatures, some fish have developed the ability to remain functional even when the water temperature is below freezing; some use natural antifreeze or antifreeze proteins to resist ice crystal formation in their tissues.

Most sharks are "cold-blooded" or, more precisely, poikilothermic, meaning that their internal body temperature matches that of their ambient environment. Members of the family Lamnidae (such as the shortfin mako shark and the great white shark) are homeothermic and maintain a higher body temperature than the surrounding water. In these sharks, a strip of aerobic red muscle located near the center of the body generates the heat, which the body retains via a countercurrent exchange mechanism by a system of blood vessels called the rete mirabile ("miraculous net"). The common thresher shark has a similar mechanism for maintaining an elevated body temperature, which is thought to have evolved independently.

Tuna can maintain the temperature of certain parts of their body above the temperature of ambient seawater. For example, bluefin tuna maintain a core body temperature of 25–33°C (77–91°F), in water as cold as 6°C (43°F). However, unlike typical endothermic creatures such as mammals and birds, tuna do not maintain temperature within a relatively narrow range. Tuna achieve endothermy by conserving the heat generated through normal metabolism. The rete mirabile ("wonderful net"), the intertwining of veins and arteries in the body's periphery, transfers heat from venous blood to arterial blood via a counter-current exchange system, thus mitigating the effects of surface cooling. This allows the tuna to elevate the temperatures of the highly aerobic tissues of the skeletal muscles, eyes and brain, which supports faster swimming speeds and reduced energy expenditure, and which enables them to survive in cooler waters over a wider range of ocean environments than those of other fish. In all tunas, however, the heart operates at ambient temperature, as it receives cooled blood, and coronary circulation is directly from the gills.

- Homeothermy: Although most fish are exclusively ectothermic, there are exceptions. Certain species of fish maintain elevated body temperatures. Endothermic teleosts (bony fish) are all in the suborder Scombroidei and include the billfishes, tunas, including a "primitive" mackerel species, *Gasterochisma melampus*. All sharks in the family Lamnidae – shortfin mako, long fin mako, white, porbeagle, and salmon shark – are endothermic, and evidence suggests the trait exists in family Alopiidae (thresher sharks). The degree of endothermy varies from the billfish, which warm only their eyes and brain, to bluefin tuna and porbeagle sharks who maintain body temperatures elevated in excess of 20°C above ambient water temperatures. Endothermy, though metabolically costly, is thought to provide advantages such as increased muscle strength, higher rates of central nervous system processing, and higher rates of digestion.

In some fish, a rete mirabile allows for an increase in muscle temperature in regions where this network of vein and arteries is found. The fish is able to thermoregulate certain areas of their body. Additionally, this increase in temperature leads to an increase in basal metabolic temperature. The fish is now able to split ATP at a higher rate and ultimately can swim faster.

The eye of a swordfish can generate heat to better cope with detecting their prey at depths of 2000 feet.

Muscular System

Fish swim by contracting longitudinal red muscle and obliquely oriented white muscles. The red muscle is aerobic and needs oxygen which is supplied by myoglobin. The white muscle is anaerobic and it does not need oxygen. Red muscles are used for sustained activity such as cruising at slow speeds on ocean migrations. White muscles are used for bursts of activity, such as jumping or sudden bursts of speed for catching prey.

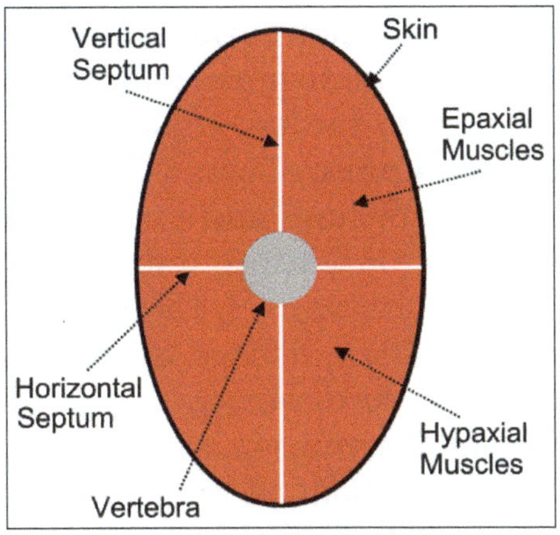

Cross section of a teleost.

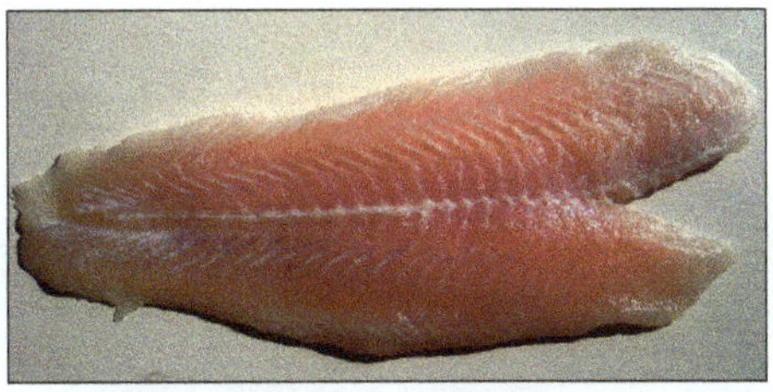

Iridescent shark filets showing myomere structure.

Mostly fish have white muscles, but the muscles of some fishes, such as scombroids and salmonids, range from pink to dark red. The red myotomal muscles derive their colour from myoglobin, an oxygen-binding molecule, which tuna express in quantities far higher than most other fish. The oxygen-rich blood further enables energy delivery to their muscles.

Most fish move by alternately contracting paired sets of muscles on either side of the backbone. These contractions form S-shaped curves that move down the body. As each curve reaches the back fin, backward force is applied to the water, and in conjunction with the fins, moves the fish forward. The fish's fins function like an airplane's flaps. Fins also increase the tail's surface area, increasing speed. The streamlined body of the fish decreases the amount of friction from the water.

A typical characteristic of many animals that utilize undulatory locomotion is that they have segmented muscles, or blocks of myomeres, running from their head to tails which are separated by connective tissue called myosepta. In addition, some segmented muscle groups, such the lateral hypaxial musculature in the salamander are oriented at an angle to the longitudinal direction. For these obliquely oriented fibers the strain in the longitudinal direction is greater than the strain in the muscle fiber direction leading to an architectural gear ratio greater than 1. A higher initial angle of orientation and more dorsoventral bulging produces a faster muscle contraction but results in a lower amount of force production. It is hypothesized that animals employ a variable gearing mechanism that allows self-regulation of force and velocity to meet the mechanical demands of the contraction. When a pennate muscle is subjected to a low force, resistance to width changes in the muscle cause it to rotate which consequently produce a higher architectural gear ratio (AGR) (high velocity). However, when subject to a high force, the perpendicular fiber force component overcomes the resistance to width changes and the muscle compresses producing a lower AGR (capable of maintaining a higher force output).

Most fishes bend as a simple, homogenous beam during swimming via contractions of longitudinal red muscle fibers and obliquely oriented white muscle fibers within the segmented axial musculature. The fiber strain (εf) experienced by the longitudinal red muscle fibers is equivalent to the longitudinal strain (εx). The deeper white muscle fibers fishes show diversity in arrangement. These fibers are organized into cone-shaped structures and attach to connective tissue sheets known as myosepta; each fiber shows a characteristic dorsoventral (α) and mediolateral (φ) trajectory. The segmented architecture theory predicts that, εx > εf. This phenomenon results in an architectural gear ratio, determined as longitudinal strain divided by fiber strain (εx / εf), greater than one and longitudinal velocity amplification; furthermore, this emergent velocity amplification may be augmented by variable architectural gearing via mesolateral and dorsoventral shape changes, a pattern seen in pennate muscle contractions. A red-to-white gearing ratio (red εf / white εf) captures the combined effect of the longitudinal red muscle fiber and oblique white muscle fiber strains.

Buoyancy

Swim bladder of a common rudd.

The body of a fish is denser than water, so fish must compensate for the difference or they will sink. Many bony fishes have an internal organ called a swim bladder, or gas bladder, that adjusts their buoyancy through manipulation of gases. In this way, fish can stay at the current water depth, or ascend or descend without having to waste energy in swimming. The bladder is only found in bony fishes. In the more primitive groups like some minnows, bichirs and lungfish, the bladder is open to the esophagus and double as a lung. It is often absent in fast swimming fishes such as the tuna and mackerel families. The condition of a bladder open to the esophagus is called physostome, the closed condition physoclist. In the latter, the gas content of the bladder is controlled through the rete mirabilis, a network of blood vessels effecting gas exchange between the bladder and the blood.

Sharks, like this three tonne great white shark, don't have swim bladders. Most sharks need to keep swimming to avoid sinking.

In some fish, a rete mirabile fills the swim bladder with oxygen. A countercurrent exchange system is utilized between the venous and arterial capillaries. By lowering the pH levels in the venous capillaries, oxygen unbinds from blood hemoglobin. This causes an increase in venous blood oxygen concentration, allowing the oxygen to diffuse through the capillary membrane and into the arterial capillaries, where oxygen is still sequestered to hemoglobin. The cycle of diffusion continues until the concentration of oxygen in the arterial capillaries is supersaturated (larger than the concentration of oxygen in the swim bladder). At this point, the free oxygen in the arterial capillaries diffuses into the swim bladder via the gas gland.

Unlike bony fish, sharks do not have gas-filled swim bladders for buoyancy. Instead, sharks rely on a large liver filled with oil that contains squalene, and their cartilage, which is about half the normal density of bone. Their liver constitutes up to 30% of their total body mass. The liver's effectiveness is limited, so sharks employ dynamic lift to maintain depth when not swimming. Sand tiger sharks store air in their stomachs, using it as a form of swim bladder. Most sharks need to constantly swim in order to breathe and cannot sleep very long without sinking (if at all). However, certain species, like the nurse shark, are capable of pumping water across their gills, allowing them to rest on the ocean bottom.

Sensory Systems

Most fish possess highly developed sense organs. Nearly all daylight fish have color vision that is at least as good as a human's (see vision in fishes). Many fish also have chemoreceptors that are responsible for extraordinary senses of taste and smell. Although they have ears, many fish may not hear very well. Most fish have sensitive receptors that form the lateral line system, which detects gentle currents and vibrations, and senses the motion of nearby fish and prey. Sharks can sense frequencies in the range of 25 to 50 Hz through their lateral line.

Fish orient themselves using landmarks and may use mental maps based on multiple landmarks or symbols. Fish behavior in mazes reveals that they possess spatial memory and visual discrimination.

Vision

Vision is an important sensory system for most species of fish. Fish eyes are similar to those of terrestrial vertebrates like birds and mammals, but have a more spherical lens. Their retinas generally have both rod cells and cone cells (for scotopic and photopic vision), and most species have colour vision. Some fish can see ultraviolet and some can see polarized light. Amongst jawless fish, the lamprey has well-developed eyes, while the hagfish has only primitive eyespots. Fish vision shows adaptation to their visual environment, for example deep sea fishes have eyes suited to the dark environment.

Hearing

Hearing is an important sensory system for most species of fish. Hearing threshold and the ability to localize sound sources are reduced underwater, in which the speed of sound is faster than in air. Underwater hearing is by bone conduction, and localization of sound appears to depend on differences in amplitude detected by bone conduction. Aquatic animals such as fish, however, have a more specialized hearing apparatus that is effective underwater.

Fish can sense sound through their lateral lines and their otoliths (ears). Some fishes, such as some species of carp and herring, hear through their swim bladders, which function rather like a hearing aid.

Hearing is well-developed in carp, which have the Weberian organ, three specialized vertebral processes that transfer vibrations in the swim bladder to the inner ear.

Although it is hard to test sharks' hearing, they may have a sharp sense of hearing and can possibly hear prey many miles away. A small opening on each side of their heads (not the spiracle) leads directly into the inner ear through a thin channel. The lateral line shows a similar arrangement, and is open to the environment via a series of openings called lateral line pores. This is a reminder of the common origin of these two vibration- and sound-detecting organs that are grouped together as the acoustico-lateralis system. In bony fish and tetrapods the external opening into the inner ear has been lost.

Chemoreception

The shape of the hammerhead shark's head may enhance olfaction by spacing the nostrils further apart.

Sharks have keen olfactory senses, located in the short duct (which is not fused, unlike bony fish) between the anterior and posterior nasal openings, with some species able to detect as little as one part per million of blood in seawater.

Sharks have the ability to determine the direction of a given scent based on the timing of scent detection in each nostril. This is similar to the method mammals use to determine direction of sound.

They are more attracted to the chemicals found in the intestines of many species, and as a result often linger near or in sewage outfalls. Some species, such as nurse sharks, have external barbels that greatly increase their ability to sense prey.

Magnetoception

Electroreception

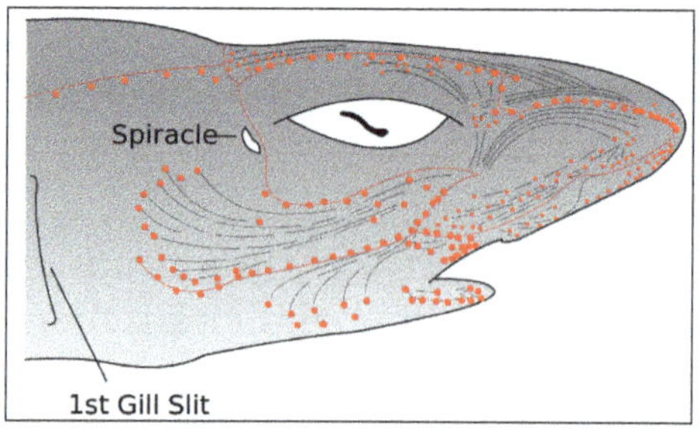

Electromagnetic field receptors (ampullae of Lorenzini) and motion detecting canals in the head of a shark.

Some fish, such as catfish and sharks, have organs that detect weak electric currents on the order of millivolt. Other fish, like the South American electric fishes Gymnotiformes, can produce weak electric currents, which they use in navigation and social communication. In sharks, the ampullae of Lorenzini are electroreceptor organs. They number in the hundreds to thousands. Sharks use the ampullae of Lorenzini to detect the electromagnetic fields that all living things produce. This helps sharks (particularly the hammerhead shark) find prey. The shark has the greatest electrical sensitivity of any animal. Sharks find prey hidden in sand by detecting the electric fields they produce. Ocean currents moving in the magnetic field of the Earth also generate electric fields that sharks can use for orientation and possibly navigation.

- The ampullae of Lorenzini allow sharks to sense electrical discharges.
- Electric fish are able to produce electric fields by modified muscles in their body.

Pain

Experiments done by William Tavolga provide evidence that fish have pain and fear responses. For instance, in Tavolga's experiments, toadfish grunted when electrically shocked and over time they came to grunt at the mere sight of an electrode.

In 2003, Scottish scientists at the University of Edinburgh and the Roslin Institute concluded that rainbow trout exhibit behaviors often associated with pain in other animals. Bee venom and acetic acid injected into the lips resulted in fish rocking their bodies and rubbing their lips along the sides and floors of their tanks, which the researchers concluded were attempts to relieve pain, similar to what mammals would do. Neurons fired in a pattern resembling human neuronal patterns.

Professor James D. Rose of the University of Wyoming claimed the study was flawed since it did not provide proof that fish possess "conscious awareness, particularly a kind of awareness that is meaningfully like ours". Rose argues that since fish brains are so different from human brains, fish are probably not conscious in the manner humans are, so that reactions similar to human reactions to pain instead have other causes. Rose had published a study a year earlier arguing that fish cannot feel pain because their brains lack a neocortex. However, animal behaviorist Temple Grandin argues that fish could still have consciousness without a neocortex because "different species can use different brain structures and systems to handle the same functions."

Animal welfare advocates raise concerns about the possible suffering of fish caused by angling. Some countries, such as Germany have banned specific types of fishing, and the British RSPCA now formally prosecutes individuals who are cruel to fish.

Reproductive Processes

Oogonia development in teleosts fish varies according to the group, and the determination of oogenesis dynamics allows the understanding of maturation and fertilisation processes. Changes in the nucleus, ooplasm, and the surrounding layers characterize the oocyte maturation process.

Postovulatory follicles are structures formed after oocyte release; they do not have endocrine function, present a wide irregular lumen, and are rapidly reabsorbed in a process involving the apoptosis of follicular cells. A degenerative process called follicular atresia reabsorbs vitellogenic oocytes not spawned. This process can also occur, but less frequently, in oocytes in other development stages.

Some fish are hermaphrodites, having both testes and ovaries either at different phases in their life cycle or, as in hamlets, have them simultaneously.

Over 97% of all known fish are oviparous, that is, the eggs develop outside the mother's body. Examples of oviparous fish include salmon, goldfish, cichlids, tuna, and eels. In the majority of these species, fertilisation takes place outside the mother's body, with the male and female fish shedding their gametes into the surrounding water. However, a few oviparous fish practice internal fertilisation, with the male using some sort of intromittent organ to deliver sperm into the genital opening of the female, most notably the oviparous sharks, such as the horn shark, and oviparous rays, such as skates. In these cases, the male is equipped with a pair of modified pelvic fins known as claspers.

Marine fish can produce high numbers of eggs which are often released into the open water column. The eggs have an average diameter of 1 millimetre (0.039 in). The eggs are generally surrounded by the extraembryonic membranes but do not develop a shell, hard or soft, around these membranes. Some fish have thick, leathery coats, especially if they must withstand physical force or desiccation. These type of eggs can also be very small and fragile.

Egg of lamprey.

Egg of catshark (mermaids' purse).

Egg of bullhead shark.

Egg of chimaera.

The newly hatched young of oviparous fish are called larvae. They are usually poorly formed, carry a large yolk sac (for nourishment) and are very different in appearance from juvenile and adult specimens. The larval period in oviparous fish is relatively short (usually only several weeks), and larvae rapidly grow and change appearance and structure (a process termed metamorphosis) to become juveniles. During this transition larvae must switch from their yolk sac to feeding on zooplankton prey, a process which depends on typically inadequate zooplankton density, starving many larvae.

In ovoviviparous fish the eggs develop inside the mother's body after internal fertilisation but receive little or no nourishment directly from the mother, depending instead on the yolk. Each embryo develops in its own egg. Familiar examples of ovoviviparous fish include guppies, angel sharks, and coelacanths.

Some species of fish are viviparous. In such species the mother retains the eggs and nourishes the embryos. Typically, viviparous fish have a structure analogous to the placenta seen in mammals connecting the mother's blood supply with that of the embryo. Examples of viviparous fish include the surf-perches, splitfins, and lemon shark. Some viviparous fish exhibit oophagy, in which the developing embryos eat other eggs produced by the mother. This has been observed primarily among sharks, such as the shortfin mako and porbeagle, but is known for a few bony fish as well, such as the halfbeak *Nomorhamphus ebrardtii*. Intrauterine cannibalism is an even more unusual mode of vivipary, in which the largest embryos eat weaker and smaller siblings. This behavior is also most commonly found among sharks, such as the grey nurse shark, but has also been reported for *Nomorhamphus ebrardtii*.

In many species of fish, fins have been modified to allow Internal fertilisation.

Aquarists commonly refer to ovoviviparous and viviparous fish as livebearers.

- Many fish species are hermaphrodites. *Synchronous hermaphrodites* possess both ovaries and testes at the same time. *Sequential hermaphrodites* have both types of tissue in their gonads, with one type being predominant while the fish belongs to the corresponding gender.

Fish Jaw

Most bony fishes have two sets of jaws made mainly of bone. The primary oral jaws open and close the mouth, and a second set of pharyngeal jaws are positioned at the back of the throat. The oral jaws are used to capture and manipulate prey by biting and crushing. The pharyngeal jaws, so-called because they are positioned within the pharynx, are used to further process the food and move it from the mouth to the stomach.

Cartilaginous fishes, such as sharks and rays, have one set of oral jaws made mainly of cartilage. They do not have pharyngeal jaws. Generally jaws are articulated and oppose vertically, comprising an upper jaw and a lower jaw and can bear numerous ordered teeth. Bony fishes usually develop only one set of teeth *(monophyodont)*. Cartilaginous fishes grow multiple sets *(polyphyodont)* and replace teeth as they wear.

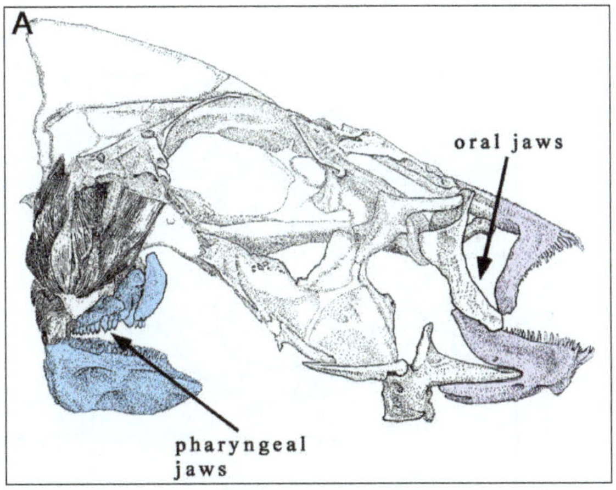

Skull of a generalized cichlid, showing a lateral view of the oral jaws (purple) and the pharyngeal jaws (blue).

Jaws probably originated in the pharyngeal arches supporting the gills of jawless fish. The earliest jaws appeared in the now extinct placoderms and spiny sharks during the Silurian, about 430 million years ago. The original selective advantage offered by the jaw was probably not related to feeding, but to increased respiration efficiency — the jaws were used in the buccal pump to pump water across the gills. The familiar use of jaws for feeding would then have developed as a secondary function before becoming the primary function in many vertebrates. All vertebrate jaws, including the human jaw, evolved from early fish jaws. The appearance of the early vertebrate jaw has been described as "perhaps the most profound and radical evolutionary step in the vertebrate history". Fish without jaws had more difficulty surviving than fish with jaws, and most jawless fish became extinct.

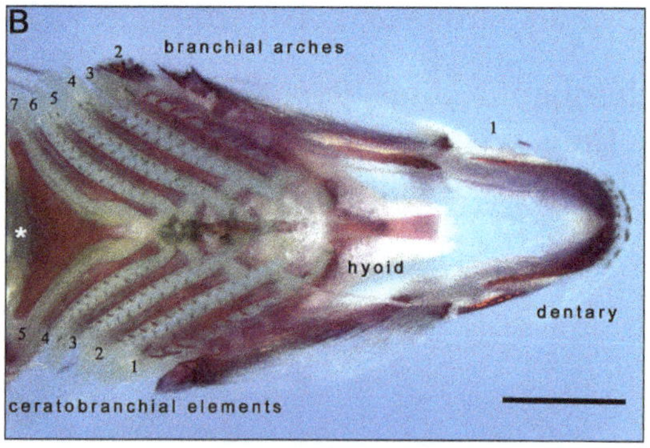

Dorsal view of the lower pharyngeal and oral jaws of a juvenile Malawi eyebiter showing the branchial (pharyngeal) arches and ceratobrachial elements (arch bones). The white asterisk indicates the toothed pharyngeal jaw. Scale bar represents 500 μm.

Jaws use linkage mechanisms. These linkages can be especially common and complex in the head of bony fishes, such as wrasses, which have evolved many specialized feeding mechanisms. Especially advanced are the linkage mechanisms of jaw protrusion. For suction feeding a system of linked four-bar linkages is responsible for the coordinated opening of the mouth and the three-dimensional expansion of the buccal cavity. Other linkages are responsible for protrusion of the premaxilla. Linkage systems are widely distributed in animals. The most thorough overview of the different types of linkages in animals has been provided by M. Muller, who also designed a new classification system, which is especially well suited for biological systems.

Skull

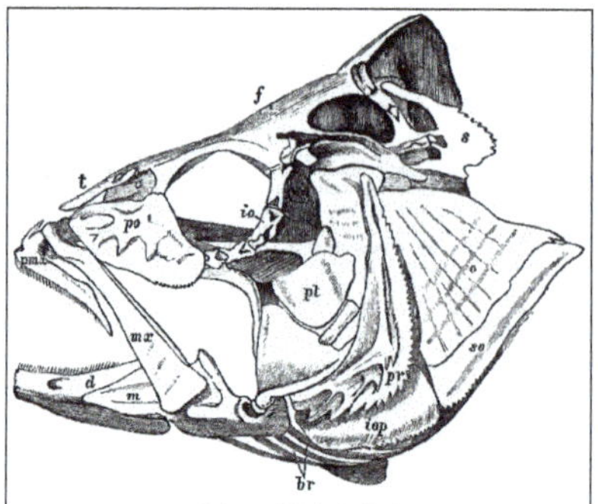

Skeleton of Head of a Perch.

f, frontal.
t, turbinal.
po, preorbital.
io, infraorbital ring.
mx, maxillary.
pmx, premaxillary.
m, mandible.
d, dentary bone.
pt, posttympanic.
s, suprascapula.
o, opercle.
so, subopercle.
pr, preopercle.
iop, interopercle.
br, branchiostegal rays.

The skull of fishes is formed from a series of loosely connected bones. Lampreys and sharks only possess a cartilaginous endocranium, with both the upper and lower jaws being separate elements.

Bony fishes have additional dermal bone, forming a more or less coherent skull roof in lungfish and holost fish. The lower jaw defines a chin.

The simpler structure is found in jawless fish, in which the cranium is represented by a trough-like basket of cartilaginous elements only partially enclosing the brain, and associated with the capsules for the inner ears and the single nostril. Distinctively, these fish have no jaws.

Cartilaginous fish, such as sharks, also have simple skulls. The cranium is a single structure forming a case around the brain, enclosing the lower surface and the sides, but always at least partially open at the top as a large fontanelle. The most anterior part of the cranium includes a forward plate of cartilage, the rostrum, and capsules to enclose the olfactory organs. Behind these are the orbits, and then an additional pair of capsules enclosing the structure of the inner ear. Finally, the skull tapers towards the rear, where the foramen magnum lies immediately above a single condyle, articulating with the first vertebra. There are, in addition, at various points throughout the cranium, smaller foramina for the cranial nerves. The jaws consist of separate hoops of cartilage, almost always distinct from the cranium proper.

In ray-finned fishes, there has also been considerable modification from the primitive pattern. The roof of the skull is generally well formed, and although the exact relationship of its bones to those of tetrapods is unclear, they are usually given similar names for convenience. Other elements of the skull, however, may be reduced; there is little cheek region behind the enlarged orbits, and little, if any bone in between them. The upper jaw is often formed largely from the premaxilla, with the maxilla itself located further back, and an additional bone, the symplectic, linking the jaw to the rest of the cranium.

Although the skulls of fossil lobe-finned fish resemble those of the early tetrapods, the same cannot be said of those of the living lungfishes. The skull roof is not fully formed, and consists of multiple, somewhat irregularly shaped bones with no direct relationship to those of tetrapods. The upper jaw is formed from the pterygoids and vomers alone, all of which bear teeth. Much of the skull is formed from cartilage, and its overall structure is reduced.

Oral Jaws

Lower

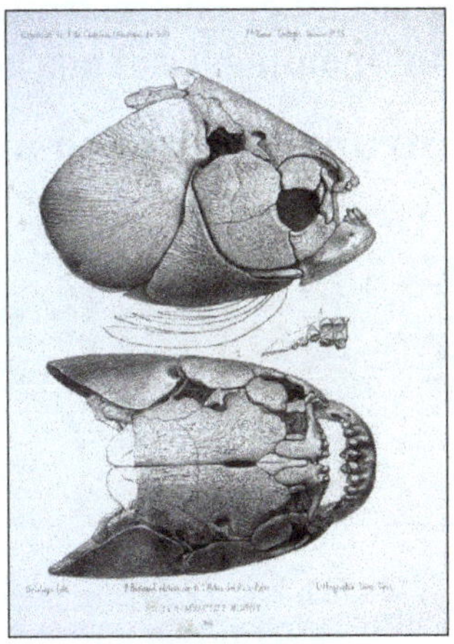

Oral jaw from side and above of *Piaractus brachypomus*, a close relative of piranhas.

In vertebrates, the lower jaw (mandible or jawbone) is a bone forming the skull with the cranium. In lobe-finned fishes and the early fossil tetrapods, the bone homologous to the mandible of mammals is merely the largest of several bones in the lower jaw. It is referred to as the *dentary bone*, and forms the body of the outer surface of the jaw. It is bordered below by a number of splenial bones, while the angle of the jaw is formed by a lower angular bone and a suprangular bone just above it. The inner surface of the jaw is lined by a *prearticular* bone, while the articular bone forms the articulation with the skull proper. Finally a set of three narrow *coronoid bones* lie above the prearticular bone. As the name implies, the majority of the teeth are attached to the dentary, but there are commonly also teeth on the coronoid bones, and sometimes on the prearticular as well.

This complex primitive pattern has, however, been simplified to various degrees in the great majority of vertebrates, as bones have either fused or vanished entirely. In teleosts, only the dentary, articular, and angular bones remain. Cartilagenous fish, such as sharks, do not have any of the bones found in the lower jaw of other vertebrates. Instead, their lower jaw is composed of a cartilagenous structure homologous with the Meckel's cartilage of other groups. This also remains a significant element of the jaw in some primitive bony fish, such as sturgeons.

Upper

The upper jaw, or maxilla is a fusion of two bones along the palatal fissure that form the upper jaw. This is similar to the mandible (lower jaw), which is also a fusion of two halves at the mandibular symphysis. In bony fish, the maxilla is called the "upper maxilla," with the mandible being the "lower maxilla". The alveolar process of the maxilla holds the upper teeth, and is referred to as the maxillary arch. In most vertebrates, the foremost part of the upper jaw, to which the incisors are attached in mammals consists of a separate pair of bones, the premaxillae. In bony fish, both maxilla and premaxilla are relatively plate-like bones, forming only the sides of the upper jaw, and part of the face, with the premaxilla also forming the lower boundary of the nostrils. Cartilaginous fish, such as sharks and rays also lack a true maxilla. Their upper jaw is instead formed from a cartilagenous bar that is not homologous with the bone found in other vertebrates.

Some fish have permanently protruding upper jawbones called rostrums. Billfish (marlin, swordfish and sailfish) use rostrums (bills) to slash and stun prey. Paddlefish, goblin sharks and hammerhead sharks have rostrums packed with electroreceptors which signal the presence of prey by detecting weak electrical fields. Sawsharks and the critically endangered sawfish have rostrums (saws) which are both electro-sensitive and used for slashing. The rostrums extend ventrally in front of the fish. In the case of hammerheads the rostrum (hammer) extends both ventrally and laterally (sideways).

Fish with rostrums (extended upper jawbones)

Sailfish, like all billfish, have a rostrum (bill) which evolved from the upper jawbone.

The paddlefish has a rostrum packed with electroreceptors.

Sawfish have an electro-sensitive rostrum (saw) which is also used to slash at prey.

Pharyngeal Jaws

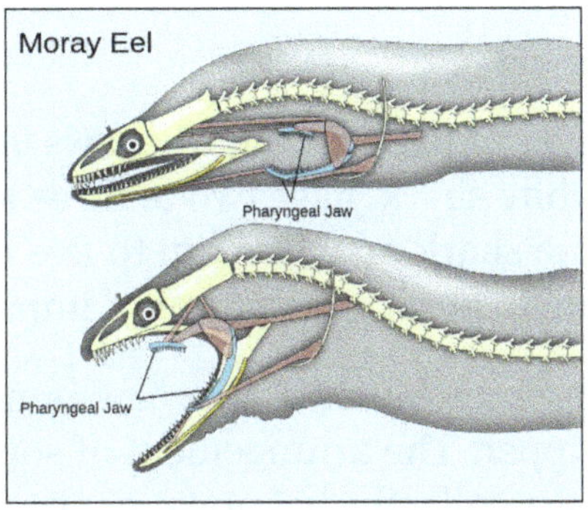

Moray eels have two sets of jaws: the oral jaws that capture prey and the pharyngeal jaws that advance into the mouth and move prey from the oral jaws to the esophagus for swallowing.

Pharyngeal jaws are a second set of jaws distinct from the primary (oral) jaws. They are contained within the throat, or pharynx, of most bony fish. They are believed to have originated, in a similar way to oral jaws, as a modification of the fifth gill arch which no longer has a respiratory function. The first four arches still function as gills. Unlike the oral jaw, the pharyngeal jaw has no jaw joint, but is supported instead by a sling of muscles.

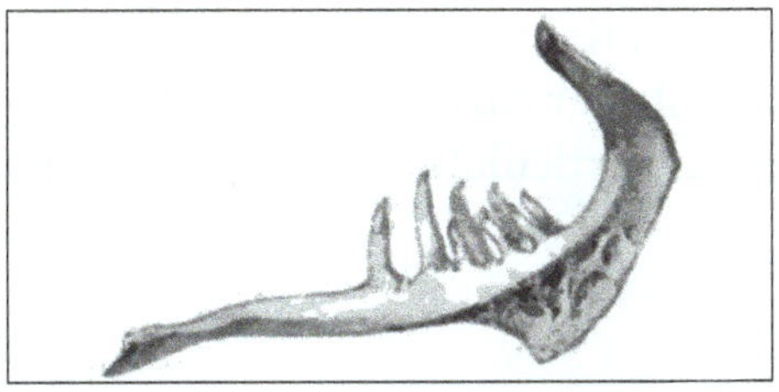

Pharyngeal jaw of an asp carrying some pharyngeal teeth.

A notable example occurs with the moray eel. The pharyngeal jaws of most fishes are not mobile. The pharyngeal jaws of the moray are highly mobile, perhaps as an adaptation to the constricted nature of the burrows they inhabit which inhibits their ability to swallow as other fishes do by creating a negative pressure in the mouth. Instead, when the moray bites prey, it first bites normally with its oral jaws, capturing the prey. Immediately thereafter, the pharyngeal jaws are brought forward and bite down on the prey to grip it; they then retract, pulling the prey down the moray eel's gullet, allowing it to be swallowed.

All vertebrates have a pharynx, used in both feeding and respiration. The pharynx arises during development through a series of six or more outpocketings called pharyngeal arches on the lateral sides of the head. The pharyngeal arches give rise to a number of different structures in the skeletal, muscular and circulatory systems in a manner which varies across the vertebrates. Pharyngeal arches trace back through chordates to basal deuterostomes who also share endodermal outpocketings of the pharyngeal apparatus. Similar patterns of gene expression can be detected in the developing pharynx of amphioxus and hemichordates. However, the vertebrate pharynx is unique in that it gives rise to endoskeletal support through the contribution of neural crest cells.

Cartilaginous Jaws

Cartilaginous fishes (sharks, rays and skates) have cartilaginous jaws. The jaw's surface (in comparison to the vertebrae and gill arches) needs extra strength due to its heavy exposure to physical stress. It has a layer of tiny hexagonal plates called "tesserae", which are crystal blocks of calcium salts arranged as a mosaic. This gives these areas much of the same strength found in the bony tissue found in other animals.

Generally sharks have only one layer of tesserae, but the jaws of large specimens, such as the bull shark, tiger shark, and the great white shark, have two to three layers or more, depending on body size. The jaws of a large great white shark may have up to five layers. In the rostrum (snout), the cartilage can be spongy and flexible to absorb the power of impacts.

In sharks and other extant elasmobranchs the upper jaw is not fused to the cranium, and the lower jaw is articulated with the upper. The arrangement of soft tissue and any additional articulations connecting these elements is collectively known as the jaw suspension. There are several archetypal jaw suspensions: amphistyly, orbitostyly, hyostyly, and euhyostyly. In amphistyly, the palatoquadrate has a postorbital articulation with the chondrocranium from which ligaments primarily suspend it anteriorly. The hyoid articulates with the mandibular arch posteriorly, but it appears to provide little support to the upper and lower jaws. In orbitostyly, the orbital process hinges with the orbital wall and the hyoid provides the majority of suspensory support. In contrast, hyostyly involves an ethmoid articulation between the upper jaw and the cranium, while the hyoid most likely provides vastly more jaw support compared to the anterior ligaments. Finally, in euhyostyly, also known as true hyostyly, the mandibular cartilages lack a ligamentous connection to the cranium. Instead, the hyomandibular cartilages provide the only means of jaw support, while the ceratohyal and basihyal elements articulate with the lower jaw, but are disconnected from the rest of the hyoid.

Teeth

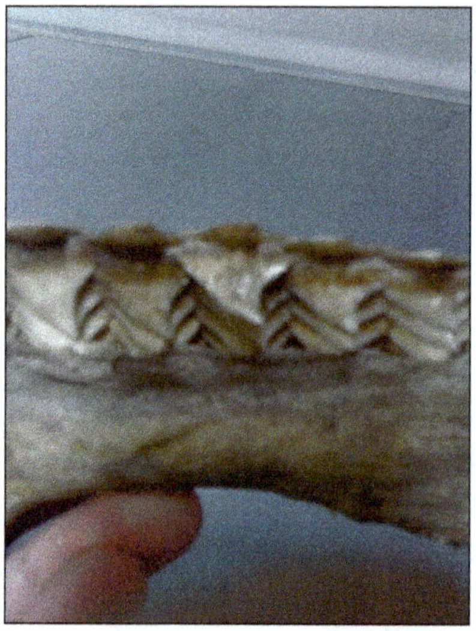

Inside of a shark jaw where new teeth move forward as though on a conveyor belt.

Jaws provide a platform in most fishes for simple pointed teeth. Lungfish and chimaera have teeth modified into broad enamel plates with jagged ridges for crushing or grinding. Carp and loach have pharyngeal teeth only. Sea horses, pipefish and adult sturgeon have no teeth of any type.

Shark teeth are embedded in the gums rather than directly affixed to the jaw, and are constantly replaced throughout life. Multiple rows of replacement teeth grow in a groove on the inside of the jaw and steadily moving forward as though on a conveyor belt. Some sharks lose 30,000 or more teeth in their lifetime. The rate of tooth replacement varies from once every 8 to 10 days to several months. In most species, teeth are replaced one at a time as opposed to the simultaneous replacement of an entire row, which is observed in the cookiecutter shark.

Tooth shape depends on the shark's diet: those that feed on mollusks and crustaceans have dense and flattened teeth used for crushing, those that feed on fish have needle-like teeth for gripping, and those that feed on larger prey such as mammals have pointed lower teeth for gripping and triangular upper teeth with serrated edges for cutting. The teeth of plankton-feeders such as the basking shark are small and non-functional.

Cartilaginous jaws and their teeth

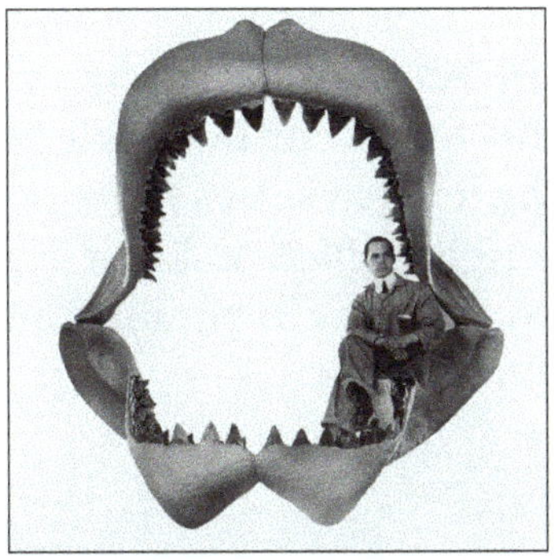

Jaw reconstruction of the extinct *Carcharodon megalodon*, 1909.

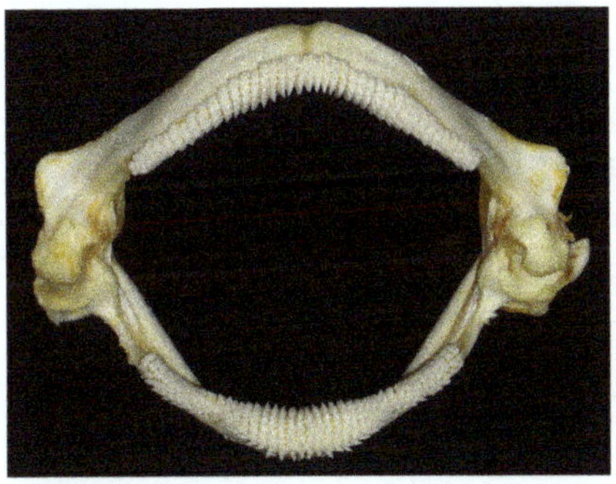

The thornback ray has teeth adapted to feed on crabs, shrimps and small fish.

Tiger shark teeth are oblique and serrated to saw through flesh.

Examples

Salmon

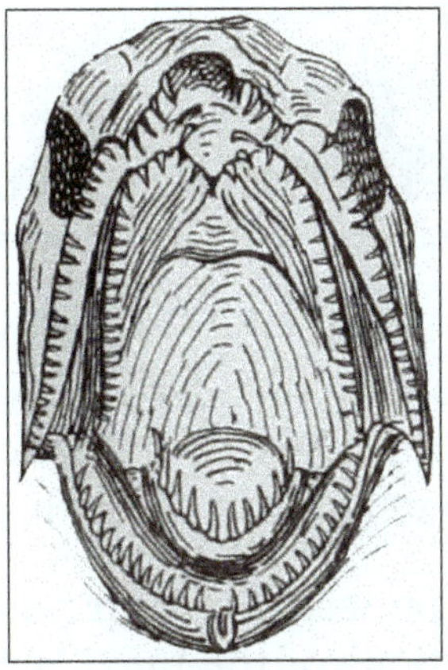

Open mouth of a salmon showing the second set of pharyngeal jaws positioned at the back of the throat.

Male salmon often remodel their jaws during spawning runs so they have a pronounced curvature. These hooked jaws are called kypes. The purpose of the kype is not altogether clear, though they can be used to establish dominance by clamping them around the base of the tail (caudal peduncle) of an opponent.

Kype of a spawning male salmon.

Cichlids

Fish jaws, like vertebrates in general, normally show bilateral symmetry. An exception occurs with the parasitic scale-eating cichlid *Perissodus microlepis*. The jaws of this fish occur in two distinct morphological forms. One morph has its jaw twisted to the left, allowing it to eat scales more readily on its victim's right flank. The other morph has its jaw twisted to the right, which makes it easier to eat scales on its victim's left flank. The relative abundance of the two morphs in populations is regulated by frequency-dependent selection.

Wrasses

Lips of a humphead wrasse.

Wrasses have become a primary study species in fish-feeding biomechanics due to their jaw structure. They have protractile mouths, usually with separate jaw teeth that jut outwards. Many species can be readily recognized by their thick lips, the inside of which is sometimes curiously folded, a peculiarity which gave rise the German name of "lip-fishes" (*Lippfische.*)

The nasal and mandibular bones are connected at their posterior ends to the rigid neurocranium, and the superior and inferior articulations of the maxilla are joined to the anterior tips of these two bones, respectively, creating a loop of 4 rigid bones connected by moving joints. This "four-bar linkage" has the property of allowing numerous arrangements to achieve a given mechanical result (fast jaw protrusion or a forceful bite), thus decoupling morphology from function. The actual morphology of wrasses reflects this, with many lineages displaying different jaw morphology that results in the same functional output in a similar or identical ecological niche.

Other

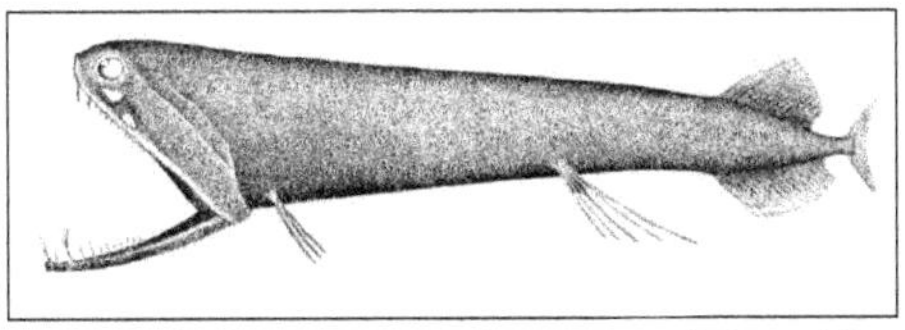

Relative to its size the stoplight loosejaw has one of the widest gapes of any fish.

Stoplight loosejaws are small fish found worldwide in the deep sea. Relative to their size they have one of the widest gapes of any fish. The lower jaw has no ethmoid membrane (floor) and is attached only by the hinge and a modified tongue bone. There are several large, fang-like teeth in the front of the jaws, followed by many small barbed teeth. There are several groups of pharyngeal teeth that serve to direct food down the esophagus.

Another deep sea fish, the pelican eel, has jaws larger than its body. The jaws are lined with small teeth and are loosely hinged. They open wide enough to swallow a fish larger than the eel itself.

Distichodontidae are a family of fresh water fishes which can be divided into genera with protractile upper jaws which are carnivores, and genera with nonprotractile upper jaws which are herbivores or predators of very small organisms.

Evolution

Vertebrate Classes

The appearance of the early vertebrate jaw has been described as "a crucial innovation" and "perhaps the most profound and radical evolutionary step in the vertebrate history". Fish without jaws had more difficulty surviving than fish with jaws, and most jawless fish became extinct. However studies of the cyclostomes, the jawless hagfishes and lampreys that did survive, have yielded little insight into the deep remodelling of the vertebrate skull that must have taken place as early jaws evolved.

The customary view is that jaws are homologous to the gill arches. In jawless fishes a series of gills opened behind the mouth, and these gills became supported by cartilaginous elements. The first set of these elements surrounded the mouth to form the jaw. The upper portion of the second embryonic arch supporting the gill became the hyomandibular bone of jawed fishes, which supports the skull and therefore links the jaw to the cranium. The hyomandibula is a set of bones found in the hyoid region in most fishes. It usually plays a role in suspending the jaws or the operculum in the case of teleosts.

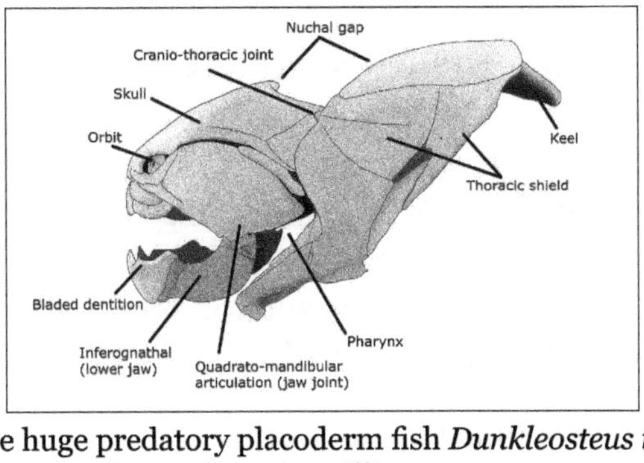

Skull diagram of the huge predatory placoderm fish *Dunkleosteus terrelli*, which lived about 380–360 million years ago.

Reconstruction of *Dunkleosteus terrelli*.

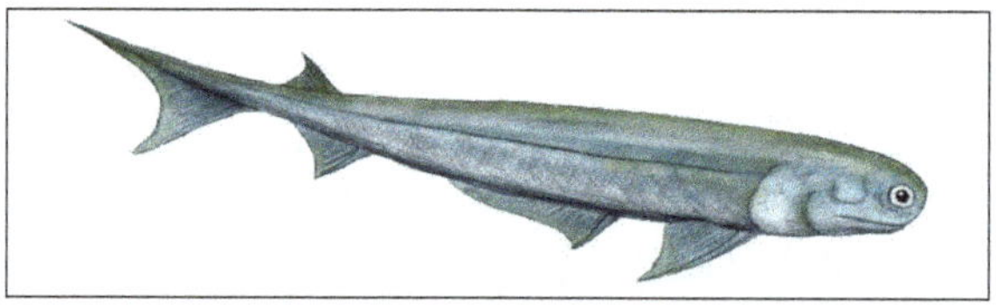

Spiny shark.

It is now accepted that the precursors of the jawed vertebrates are the long extinct bony (armoured) jawless fish, the so-called ostracoderms. The earliest known fish with jaws are the now extinct placoderms and spiny sharks.

Placoderms were a class of fish, heavily armoured at the front of their body, which first appeared in the fossil records during the Silurian about 430 million years ago. Initially they very successful, diversifying remarkably during the Devonian. They became extinct by the end of that period, about 360 million years ago. Their largest species, *Dunkleosteus terrelli*, measured up to 10 m (33 ft) and weighed 3.6 t (4.0 short tons). It possessed a four bar linkage mechanism for jaw opening that incorporated connections between the skull, the thoracic shield, the lower jaw and the jaw muscles joined together by movable joints. This mechanism allowed *Dunkleosteus terrelli* to achieve a high speed of jaw opening, opening their jaws in 20 milliseconds and completing the whole process in 50-60 milliseconds, comparable to modern fishes that use suction feeding to assist in prey capture. They could also produce high bite forces when closing the jaw, estimated at 6,000 N (1,350 lb_f) at the tip and 7,400 N (1,660 lb_f) at the blade edge in the largest individuals. The pressures generated in those regions were high enough to puncture or cut through cuticle or dermal armour suggesting that *Dunkleosteus terrelli* was perfectly adapted to prey on free-swimming, armoured prey like arthropods, ammonites, and other placoderms.

Spiny sharks were another class of fish which appeared also in the fossil records during the Silurian at about the same time as the placoderms. They were smaller than most placoderms, usually under 20 centimetres. Spiny sharks did not diversify as much as placoderms, but survived much longer into the Early Permian about 290 million years ago.

The original selective advantage offered by the jaw was not related to feeding, but to increased respiration efficiency. The jaws were used in the buccal pump still observable in modern fish and amphibians, that uses "breathing with the cheeks" to pump water across the gills of fish or air into the lungs in the case of amphibians. Over evolutionary time the more familiar use of jaws (to humans) for feeding was selected for and became a very important function in vertebrates. Many teleost fish have substantially modified jaws for suction feeding and jaw protrusion, resulting in highly complex jaws with dozens of bones involved.

Jaws are thought to derive from the pharyngeal arches that support the gills in fish. The two most anterior of these arches are thought to have become the jaw itself (see hyomandibula) and the hyoid arch, which braces the jaw against the braincase and increases mechanical efficiency. While there is no fossil evidence directly to support this theory, it makes sense in light of the numbers of pharyngeal arches that are visible in extant jawed (the Gnathostomes), which have seven arches, and primitive jawless vertebrates (the Agnatha), which have nine.

Meckel's cartilage is a piece of cartilage from which the mandibles (lower jaws) of vertebrates evolved. Originally it was the lower of two cartilages which supported the first gill arch (nearest the front) in early fish. Then it grew longer and stronger, and acquired muscles capable of closing the developing jaw. In early fish and in chondrichthyans (cartilaginous fish such as sharks), Meckel's cartilage continued to be the main component of the lower jaw. But in the adult forms of osteichthyans (bony fish) and their descendants (amphibians, reptiles, birds and mammals) the cartilage was covered in bone - although in their embryos the jaw initially develops as the Meckel's cartilage. In tetrapods the cartilage partially ossifies (changes to bone) at the rear end of the jaw and becomes the articular bone, which forms part of the jaw joint in all tetrapods except mammals.

Fish Gill

Most fish exchange gases using gills on either side of the pharynx (throat). Gills are tissues which consist of threadlike protein structures called filaments. These filaments have many functions including the transfer of ions and water, as well as the exchange of oxygen, carbon dioxide, acids and ammonia. Each filament contains a capillary network that provides a large surface area for exchanging oxygen and carbon dioxide. Fish exchange gases by pulling oxygen-rich water through their mouths and pumping it over their gills. In some fish, capillary blood flows in the opposite direction to the water, causing countercurrent exchange. The gills push the oxygen-poor water out through openings in the sides of the pharynx. Some fish, like sharks and lampreys, possess multiple gill openings. However, bony fish have a single gill opening on each side. This opening is hidden beneath a protective bony cover called an operculum.

Juvenile bichirs have external gills, a very primitive feature that they share with larval amphibians.

Breathing with Gills

Air breathing fish can be divided into *obligate* air breathers and *facultative* air breathers. Obligate air breathers, such as the African lungfish, are obligated to breathe air periodically or they suffocate. Facultative air breathers, such as the catfish *Hypostomus plecostomus*, only breathe air if they need to and can otherwise rely on their gills for oxygen. Most air breathing fish are facultative air breathers that avoid the energetic cost of rising to the surface and the fitness cost of exposure to surface predators.

All basal vertebrates breathe with gills. The gills are carried right behind the head, bordering the posterior margins of a series of openings from the esophagus to the exterior. Each gill is supported by a cartilagenous or bony gill arch. The gills of vertebrates typically develop in the walls of the pharynx, along a series of gill slits opening to the exterior. Most species employ a countercurrent exchange system to enhance the diffusion of substances in and out of the gill, with blood and water flowing in opposite directions to each other.

The gills are composed of comb-like filaments, the gill lamellae, which help increase their surface area for oxygen exchange. When a fish breathes, it draws in a mouthful of water at regular intervals. Then it draws the sides of its throat together, forcing the water through the gill openings, so that it passes over the gills to the outside. The bony fish have three pairs of arches, cartilaginous fish have five to seven pairs, while the primitive jawless fish have seven. The vertebrate ancestor no doubt had more arches, as some of their chordate relatives have more than 50 pairs of gills.

Gills usually consist of thin filaments of tissue, branches, or slender tufted processes that have a highly folded surface to increase surface area. The high surface area is crucial to the gas exchange of aquatic organisms as water contains only a small fraction of the dissolved oxygen that air does. A cubic meter of air contains about 250 grams of oxygen at STP. The concentration of oxygen in water is lower than air and it diffuses more slowly. In a litre of freshwater the oxygen content is 8 cm^3 per litre compared to 210 in the same volume of air. Water is 777 times more dense than air and is 100 times more viscous. Oxygen has a diffusion rate in air 10,000 times greater than in water. The use of sac-like lungs to remove oxygen from water would not be efficient enough to sustain life. Rather than using lungs "Gaseous exchange takes place across the surface of highly vascularised gills over which a one-way current of water is kept flowing by a specialised pumping mechanism. The density of the water prevents the gills from collapsing and lying on top of each other, which is what happens when a fish is taken out of water."

Higher vertebrates do not develop gills, the gill arches form during fetal development, and lay the basis of essential structures such as jaws, the thyroid gland, the larynx, the *columella* (corresponding to the stapes in mammals) and in mammals the malleus and incus. Fish gill slits may be the evolutionary ancestors of the tonsils, thymus gland, and Eustachian tubes, as well as many other structures derived from the embryonic branchial pouches.

Bony Fish

In bony fish, the gills lie in a branchial chamber covered by a bony operculum (*branchia* is an Ancient Greek word for gills). The great majority of bony fish species have five pairs of gills, although a few have lost some over the course of evolution. The operculum can be important in adjusting the pressure of water inside of the pharynx to allow proper ventilation of the gills, so that bony fish do not have to rely on ram ventilation (and hence near constant motion) to breathe. Valves inside the mouth keep the water from escaping.

The gill arches of bony fish typically have no septum, so that the gills alone project from the arch, supported by individual gill rays. Some species retain gill rakers. Though all but the most primitive bony fish lack a spiracle, the pseudobranch associated with it often remains, being located at the base of the operculum. This is, however, often greatly reduced, consisting of a small mass of cells without any remaining gill-like structure.

Marine teleosts also use gills to excrete electrolytes. The gills' large surface area tends to create a problem for fish that seek to regulate the osmolarity of their internal fluids. Saltwater is less dilute than these internal fluids, so saltwater fish lose large quantities of water osmotically through their gills. To regain the water, they drink large amounts of seawater and excrete the salt. Freshwater is more dilute than the internal fluids of fish, however, so freshwater fish gain water osmotically through their gills.

In some primitive bony fishes and amphibians, the larvae bear external gills, branching off from the gill arches. These are reduced in adulthood, their function taken over by the gills proper in fishes and by lungs in most amphibians. Some amphibians retain the external larval gills in adulthood, the complex internal gill system as seen in fish apparently being irrevocably lost very early in the evolution of tetrapods.

Cartilaginous Fish

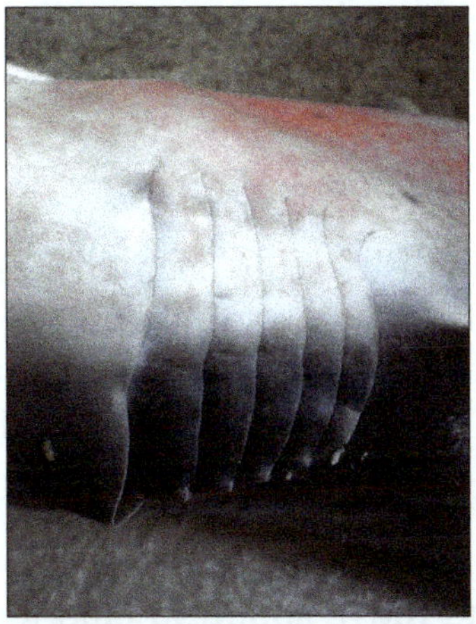

Six gill slits in a bigeyed sixgill shark; most sharks have only five.

Sharks and rays typically have five pairs of gill slits that open directly to the outside of the body, though some more primitive sharks have six or seven pairs. Adjacent slits are separated by a cartilaginous gill arch from which projects a long sheet-like septum, partly supported by a further piece of cartilage called the gill ray. The individual lamellae of the gills lie on either side of the septum. The base of the arch may also support gill rakers, small projecting elements that help to filter food from the water.

A smaller opening, the spiracle, lies in the back of the first gill slit. This bears a small pseudobranch that resembles a gill in structure, but only receives blood already oxygenated by the true gills. The spiracle is thought to be homologous to the ear opening in higher vertebrates.

Most sharks rely on ram ventilation, forcing water into the mouth and over the gills by rapidly swimming forward. In slow-moving or bottom dwelling species, especially among skates and rays, the spiracle may be enlarged, and the fish breathes by sucking water through this opening, instead of through the mouth.

Chimaeras differ from other cartilagenous fish, having lost both the spiracle and the fifth gill slit. The remaining slits are covered by an operculum, developed from the septum of the gill arch in front of the first gill.

The shared trait of breathing via gills in bony fish and cartilaginous fish is a famous example of symplesiomorphy. Bony fish are more closely related to terrestrial vertebrates, which evolved out of a clade of bony fishes that breathe through their skin or lungs, than they are to the sharks, rays, and the other cartilaginous fish. Their kind of gill respiration is shared by the "fishes" because it was present in their common ancestor and lost in the other living vertebrates. But based on this shared trait, we cannot infer that bony fish are more closely related to sharks and rays than they are to terrestrial vertebrates.

Lampreys and Hagfish

Lampreys and hagfish do not have gill slits as such. Instead, the gills are contained in spherical pouches, with a circular opening to the outside. Like the gill slits of higher fish, each pouch contains two gills. In some cases, the openings may be fused together, effectively forming an operculum. Lampreys have seven pairs of pouches, while hagfishes may have six to fourteen, depending on the species. In the hagfish, the pouches connect with the pharynx internally. In adult lampreys, a separate respiratory tube develops beneath the pharynx proper, separating food and water from respiration by closing a valve at its anterior end.

Breathing without Gills

Although most fish respire primarily using gills, some fishes can at least partially respire using mechanisms that do not require gills. In some species cutaneous respiration accounts for 5 to 40 percent of the total respiration, depending on temperature. Cutaneous respiration is more important in species that breathe air, such as mudskippers and reedfish, and in such species can account for nearly half the total respiration.

Fish from multiple groups can live out of the water for extended time periods. Amphibious fish such as the mudskipper can live and move about on land for up to several days, or live in stagnant or otherwise oxygen depleted water. Many such fish can breathe air via a variety of mechanisms. The skin of anguillid eels may absorb oxygen directly. The buccal cavity of the electric eel may breathe air. Catfish of the families Loricariidae, Callichthyidae, and Scoloplacidae absorb air through their digestive tracts. Lungfish, with the exception of the Australian lungfish, and bichirs have paired lungs similar to those of tetrapods and must surface to gulp fresh air through the mouth and pass spent air out through the gills. Gar and bowfin have a vascularized swim bladder that functions in the same way. Loaches, trahiras, and many catfish breathe by passing air through the gut. Mudskippers breathe by absorbing oxygen across the skin (similar to frogs). A number of fish have evolved so-called accessory breathing organs that extract oxygen from the air. Labyrinth fish (such as gouramis and bettas) have a labyrinth organ above the gills that performs this function. A few other fish have structures resembling labyrinth organs in form and function, most notably snakeheads, pikeheads, and the Clariidae catfish family.

Breathing air is primarily of use to fish that inhabit shallow, seasonally variable waters where the water's oxygen concentration may seasonally decline. Fish dependent solely on dissolved oxygen, such as perch and cichlids, quickly suffocate, while air-breathers survive for much longer, in some cases in water that is little more than wet mud. At the most extreme, some air-breathing fish are able to survive in damp burrows for weeks without water, entering a state of aestivation (summertime hibernation) until water returns.

Parasites on Gills

Fish gills are the preferred habitat of many ectoparasites (parasites attached to the gill but living out of it); the most commons are monogeneans and certain groups of parasitic copepods, which can be extremely numerous. Other ectoparasites found on gills are leeches and, in seawater, larvae of gnathiid isopods. Endoparasites (parasites living inside the gills) include encysted adult didymozoid trematodes, a few trichosomoidid nematodes of the genus *Huffmanela*, including *Huffmanela ossicola* which lives within the gill bone, and the encysted parasitic turbellarian *Paravortex*. Various protists and Myxosporea are also parasitic on gills, where they form cysts.

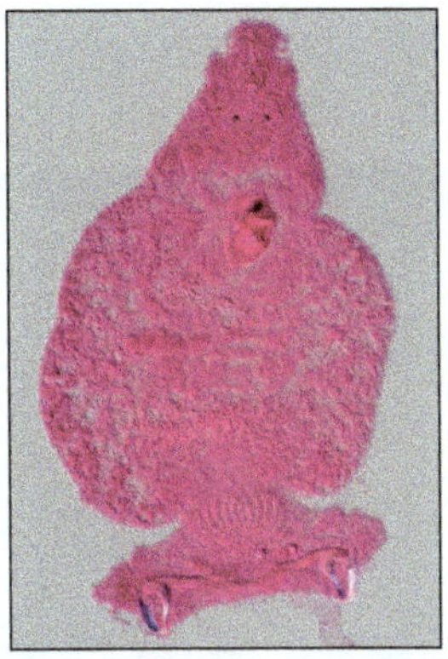

Monogenean parasite on the gill of a grouper.

Fish Scales

Cycloid scales cover these teleost fish (rohu).

The skin of most fishes are covered with scales, which, in many cases, are animal reflectors or produce animal coloration. Scales vary enormously in size, shape, structure, and extent, ranging from strong and rigid armour plates in fishes such as shrimpfishes and boxfishes, to microscopic or absent in fishes such as eels and anglerfishes. The morphology of a scale can be used to identify the species of fish it came from.

Cartilaginous fishes (sharks and rays) are covered with placoid scales. Most bony fishes are covered with the cycloid scales of salmon and carp, or the ctenoid scales of perch, or the ganoid scales of sturgeons and gars. Some species are covered instead by scutes, and others have no outer covering on the skin.

Fish scales are part of the fish's integumentary system, and are produced from the mesoderm layer of the dermis, which distinguishes them from reptile scales. The same genes involved in tooth and hair development in mammals are also involved in scale development. The placoid scales of cartilaginous fishes are also called dermal denticles and are structurally homologous with vertebrate teeth. It has been suggested that the scales of bony fishes are similar in structure to teeth, but they probably originate from different tissue. Most fish are also covered in a protective layer of mucus (slime).

Placoid Scales

Cartilaginous fishes, like this tiger shark, have placoid scales (dermal denticles).

Placoid scales are found in the cartilaginous fishes: sharks, rays, and chimaeras. They are also called *dermal denticles*. Placoid scales are structurally homologous with vertebrate teeth ("denticle" translates to "small tooth"), having a central pulp cavity supplied with blood vessels, surrounded by a conical layer of dentine, all of which sits on top of a rectangular basal plate that rests on the dermis. The outermost layer is composed of vitrodentine, a largely inorganic enamel-like substance. Placoid scales cannot grow in size, but rather more scales are added as the fish increases in size.

Similar scales can also be found under the head of the denticle herring. The amount of scale coverage is much less in rays and chimaeras.

The skin of sharks is entirely covered by placoid scales. The scales are supported by spines, which feel rough when stroked in a backward direction, but when flattened by the forward movement of water, create tiny vortices that reduce hydrodynamic drag, making swimming both more efficient as well as quieter compared to that of bony fishes. The rough, sandpaper-like texture of shark and ray skin, coupled with its toughness, has led it to be valued as a source of rawhide leather, called shagreen. One of the many historical applications of shark shagreen was in making hand-grips for swords.

Unlike bony fish, sharks have a complicated dermal corset made of flexible collagenous fibers arranged as a helical network surrounding their body. The corset works as an outer skeleton, providing attachment for their swimming muscles and thus saving energy. Their dermal teeth give them hydrodynamic advantages, as the scales reduce the turbulence of swimming.

Leptoid Scales

Leptoid scales are found on higher-order bony fish, the teleosts (the more derived clade of ray-finned fishes). As the fish grow, scales are added in concentric layers. The scales are arranged so as to overlap in a head-to-tail configuration, like roof tiles, allowing a smoother flow of water over the body and thereby reducing drag. Leptoid scales come in two forms: cycloid and ctenoid.

Cycloid Scales

Cycloid (circular) scales have a smooth texture and are uniform, with a smooth outer edge or margin. They are most common on fish with soft fin rays, such as salmon and carp.

Cycloid (Circular) Scales

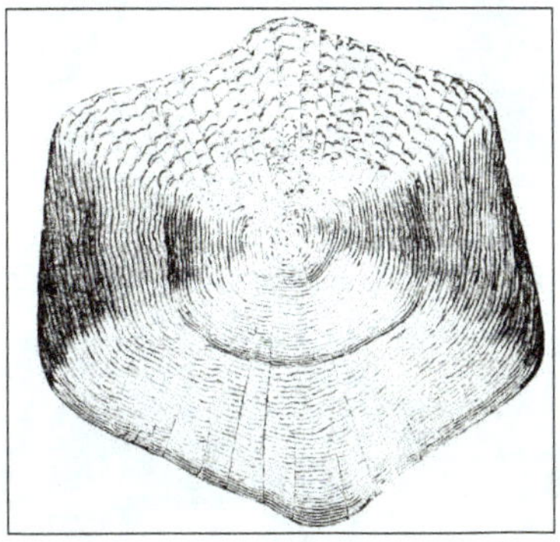

The cycloid scale of a carp has a smooth outer edge.

This *Poropuntius huguenini* is a carp-like fish with circular cycloid scales that are smooth to the touch.

Cycloid (circular) scales have a smooth texture and are uniform, with a smooth outer edge or margin. They are most common on fish with soft fin rays, such as salmon and carp.

Cycloid (circular) scales are usually found on carp-like or salmon-like fishes			
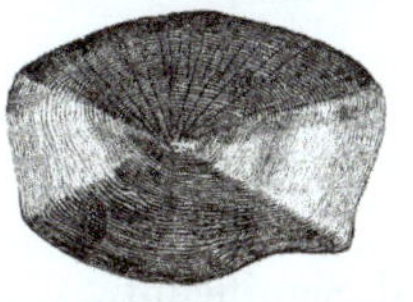	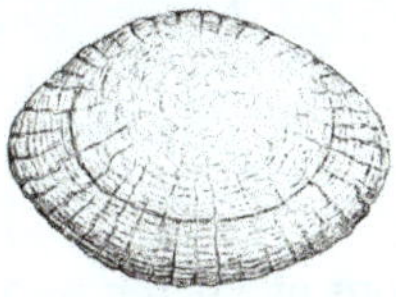	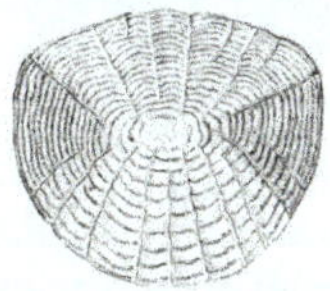	
bream	loach	minnow	grayling
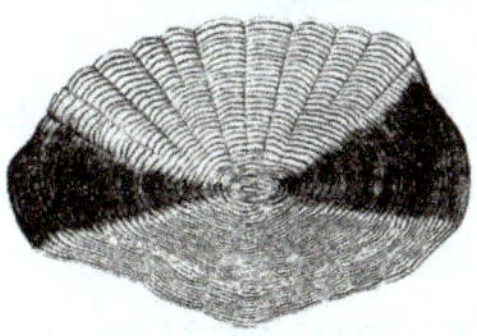	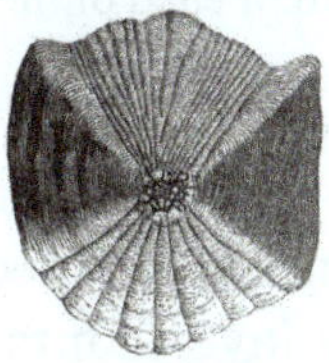	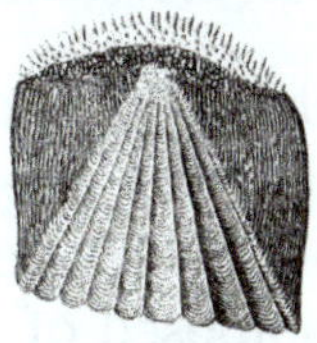	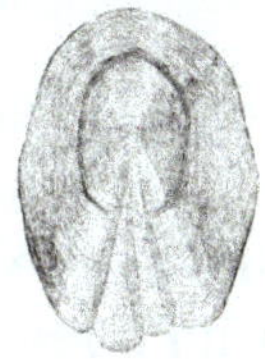
bleak	chub	gudgeon	pike

Ctenoid Scales

Ctenoid (Toothed) Scales

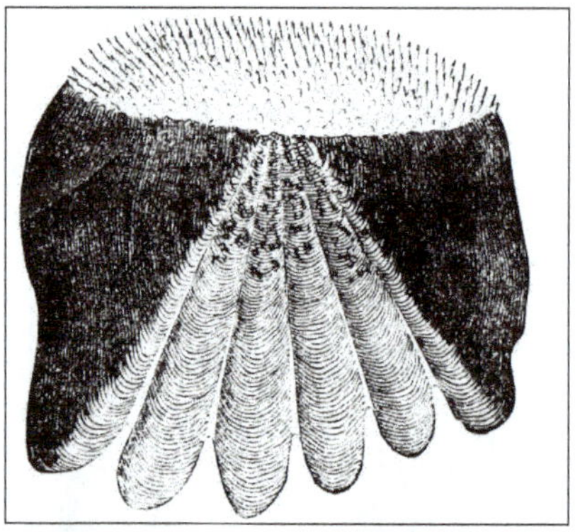

The ctenoid scale of a perch has a toothed outer edge (at top of image).

This dottyback is a perch-like fish with toothed ctenoid scales that are rough to the touch.

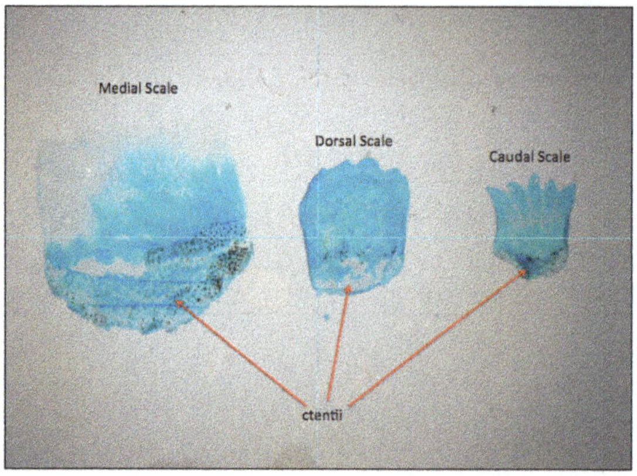

Three ctenoid scales from various locations of a perch were stained. Significant variation can be observed between the medial (middle of the fish), dorsal (top), and caudal (tail end) scales. The ctentii of each of the scales is labeled.

Ctenoid (toothed) scales are like cycloid scales, with small teeth along their outer edges. They are usually found on fishes with spiny fin rays, such as the perch-like fishes. The scales have a rough texture with a toothed outer or posterior edge featuring tiny teeth called ctenii. These scales contain almost no bone, being composed of a surface layer containing hydroxyapatite and calcium carbonate and a deeper layer composed mostly of collagen. The enamel of the other scale types is reduced to superficial ridges and ctenii.

Ctenoid (toothed) scales are usually found on perch-like fishes			
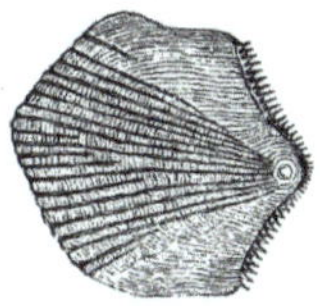	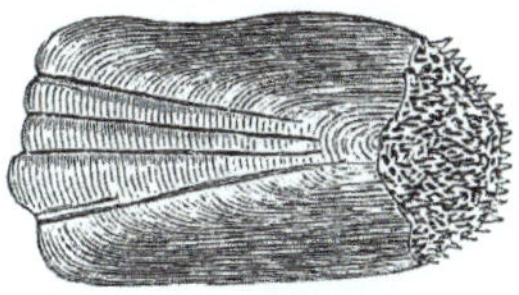	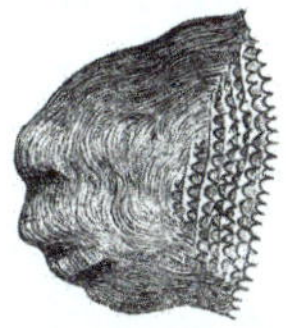	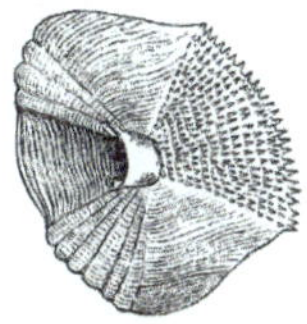
goby	flathead	scat	emperor

Ctenoid scales, similar to other epidermal structures, originate from placodes and distinctive cellular differentiation makes them exclusive from other structures that arise from the integument. Development starts near the caudal fin, along the lateral line of the fish. The development process begins with an accumulation of fibroblasts between the epidermis and dermis. Collagen fibrils begin to organize themselves in the dermal layer, which leads to the initiation of mineralization. The circumference of the scales grows first, followed by thickness when overlapping layers mineralize together.

Ctenoid scales can be further subdivided into three types:

- Crenate scales, where the margin of the scale bears indentations and projections.
- Spinoid scales, where the scale bears spines that are continuous with the scale itself.
- True ctenoid scales, where the spines on the scale are distinct structures.

Both cycloid and ctenoid scales are overlapping, making them more flexible than cosmoid and ganoid scales. Unlike ganoid scales, they grow in size through additions to the margin. The scales of some species exhibit bands of uneven seasonal growth called annuli (singular annulus). These bands can be used to age the fish. Most ray-finned fishes have ctenoid scales. Some species of flatfishes have ctenoid scales on the eyed side and cycloid scales on the blind side, while other species have ctenoid scales in males and cycloid scales in females.

Ganoid Scales

The longnose gar has diamond-shape ganoid scales.

Ganoid scales are found in the sturgeons, paddlefishes, gars, bowfin, and bichirs. They are derived from cosmoid scales, with a layer of dentine in the place of cosmine, and a layer of inorganic bone salt called ganoine in place of vitrodentine. Most are diamond-shaped and connected by peg-and-socket joints. They are usually thick and have a minimal amount of overlap as compared to other scales. In sturgeons, the scales are greatly enlarged into armour plates along the sides and back, while in the bowfin the scales are greatly reduced in thickness to resemble cycloid scales.

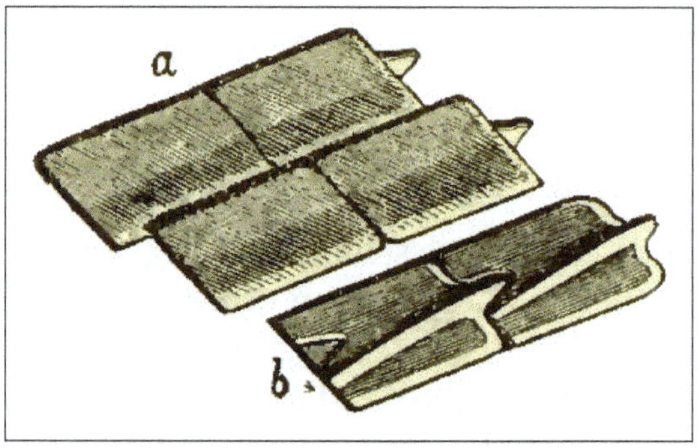

Ganoid scales of the Carboniferous fish, Amblypterus striatus. (a) shows the outer surface of four of the scales, and (b) shows the inner surface of two of the scales. Each of the rhomboidal ganoid scales of Amblypterus has a ridge on the inner surface which is produced at one end into a projecting peg which fits into a notch in the next scale, similar to the manner in which tiles are pegged together on the roof of a house.

Elasmoid Scales

Lobe-finned fishes, like this preserved coelacanth, have elasmoid scales.

Elasmoid scales are thin, imbricated scales composed of a layer of dense, lamellar bone called isopedine, above which is a layer of tubercles usually composed of bone, as in *Eusthenopteron*. The layer of dentine that was present in the first sarcopterygians is usually reduced, as in the extant coelacanth, or entirely absent, as in extant lungfish and in the Devonian *Eusthenopteron*. Elasmoid scales have appeared several times over the course of fish evolution. They are present in some lobe-finned fishes: coelacanths, all extant and some extinct lungfishes, some tetrapodomorphs like *Eusthenopteron*, amiids, and teleosts, whose cycloid and ctenoid scales represent the least mineralized elasmoid scales.

Cosmoid Scales

Cosmoid scales are found in several ancient lobe-finned fishes, including some of the earliest lungfishes, and were probably derived from a fusion of placoid scales. They are composed of a layer of dense, lamellar bone called isopedine, above which is a layer of spongy bone supplied with blood vessels. The bone layers are covered by a complex dentine layer called cosmine and a superficial outer coating of vitrodentine. Cosmoid scales increase in size through the growth of the lamellar bone layer.

Scutes

A scute is another, less common, type of scale. Scute comes from Latin for *shield*, and can take the form of:

- An external shield-like bony plate, or
- A modified, thickened scale that often is keeled or spiny, or
- A projecting, modified (rough and strongly ridged) scale, usually associated with the lateral line, or on the caudal peduncle forming caudal keels, or along the ventral profile.

Some fish, such as pineconefish, are completely or partially covered in scutes. River herrings and threadfins have an abdominal row of scutes, which are scales with raised, sharp points that are used for protection. Some jacks have a row of scutes following the lateral line on either side.

Thelodont Scales

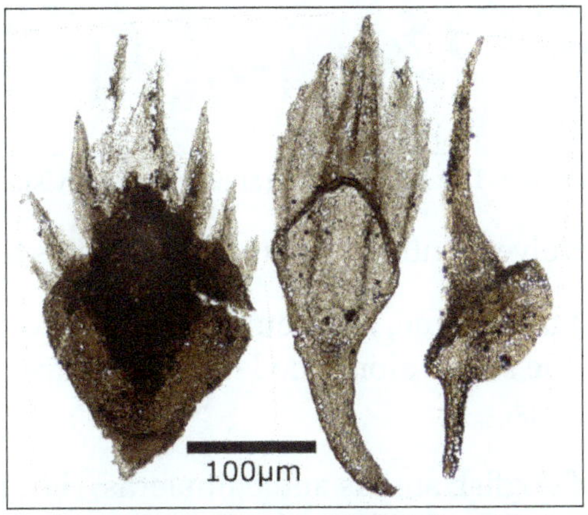

Left to right: denticles of *Paralogania* (?), *Shielia taiti*, *Lanarkia horrida*.

The bony scales of thelodonts, the most abundant form of fossil fish, are well understood. The scales were formed and shed throughout the organisms' lifetimes, and quickly separated after their death.

Bone, a tissue that is both resistant to mechanical damage and relatively prone to fossilization, often preserves internal detail, which allows the histology and growth of the scales to be studied in detail. The scales comprise a non-growing "crown" composed of dentine, with a sometimes-ornamented enameloid upper surface and an aspidine base. Its growing base is made of cell-free bone, which sometimes developed anchorage structures to fix it in the side of the fish. Beyond that, there appear to be five types of bone-growth, which may represent five natural groupings within the thelodonts—or a spectrum ranging between the end members meta- (or ortho-) dentine and mesodentine tissues. Interestingly, each of the five scale morphs appears to resemble the scales of more derived groupings of fish, suggesting that thelodont groups may have been stem groups to succeeding clades of fish.

However, using scale morphology alone to distinguish species has some pitfalls. Within each organism, scale shape varies hugely according to body area, with intermediate forms appearing between different areas—and to make matters worse, scale morphology may not even be constant within one area. To confuse things further, scale morphologies are not unique to taxa, and may be indistinguishable on the same area of two different species.

The morphology and histology of thelodonts provides the main tool for quantifying their diversity and distinguishing between species, although ultimately using such convergent traits is prone to errors. Nonetheless, a framework comprising three groups has been proposed based upon scale morphology and histology.

Modifications

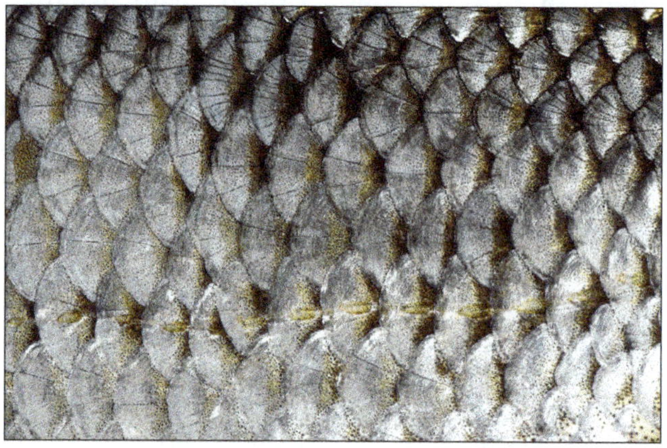

The cycloid scales of a common roach. The series of lateral line scales is visible in the lower half of the image.

Different groups of fish have evolved a number of modified scales to serve various functions.

- Almost all fishes have a lateral line, a system of mechanoreceptors that detect water movements. In bony fishes, the scales along the lateral line have central pores that allow water to contact the sensory cells.
- The dorsal fin spines of dogfish sharks and chimaeras, the stinging tail spines of stingrays, and the "saw" teeth of sawfishes and sawsharks are fused and modified placoid scales.
- Porcupine fishes have scales modified into spines.
- Surgeonfishes have a sharp, blade-like spines on either side of the caudal peduncle.
- Some herrings, anchovies, and halfbeaks have deciduous scales, which are easily shed and aid in escaping predators.
- Male Percina darters have a row of enlarged caducous scales between the pelvic fins and the anus.

Many groups of bony fishes, including pipefishes and seahorses, several families of catfishes, sticklebacks, and poachers, have developed external bony plates, structurally resembling placoid scales, as protective armour. In the boxfishes, the plates are all fused together to form a rigid shell enclosing the entire body. Yet these bony plates are not modified scales, but skin that has been ossified.

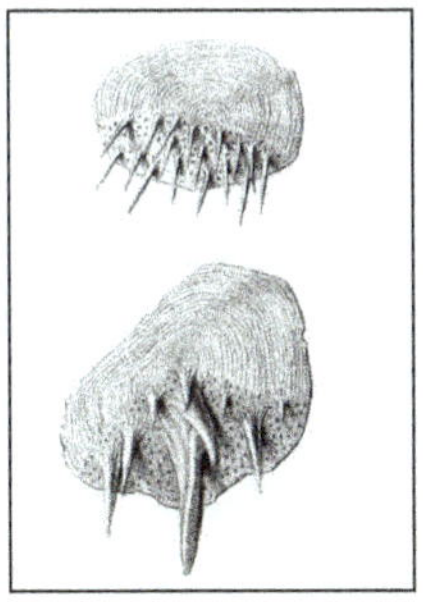
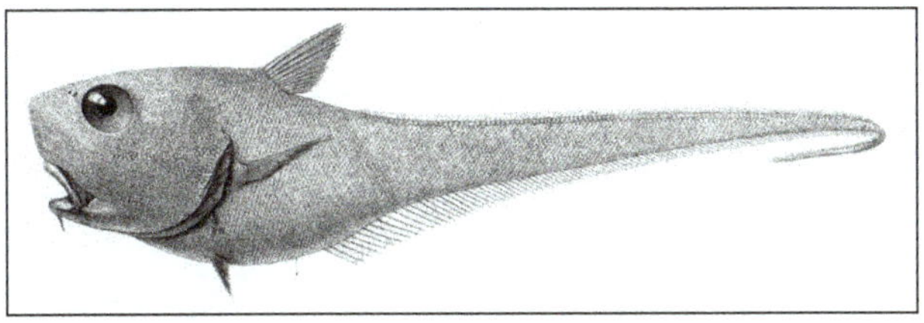

The size of the teeth on ctenoid scales can vary with position, as these scales from the rattail Cetonurus crassiceps show.

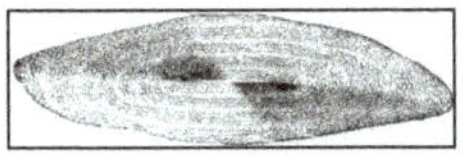

Eels seem scaleless, yet some species are covered with tiny smooth cycloid scales.

Scales of *Arapaima gigas*.

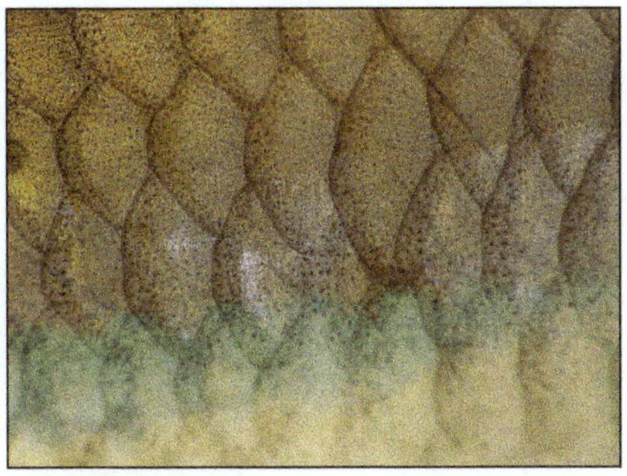

Scales of European bitterling from the hindflank.

Fish Fin

Fins are usually the most distinctive features of a fish. They are composed of bony spines or rays protruding from the body with skin covering them and joining them together, either in a webbed fashion, as seen in most bony fish, or similar to a flipper, as seen in sharks. Apart from the tail or caudal fin, fish fins have no direct connection with the spine and are supported only by muscles. Their principal function is to help the fish swim. Fins located in different places on the fish serve different purposes such as moving forward, turning, keeping an upright position or stopping. Most fish use fins when swimming, flying fish use pectoral fins for gliding, and frogfish use them for crawling. Fins can also be used for other purposes; male sharks and mosquitofish use a modified fin to deliver sperm, thresher sharks use their caudal fin to stun prey, reef stonefish have spines in their dorsal fins that inject venom, anglerfish use the first spine of their dorsal fin like a fishing rod to lure prey, and triggerfish avoid predators by squeezing into coral crevices and using spines in their fins to lock themselves in place.

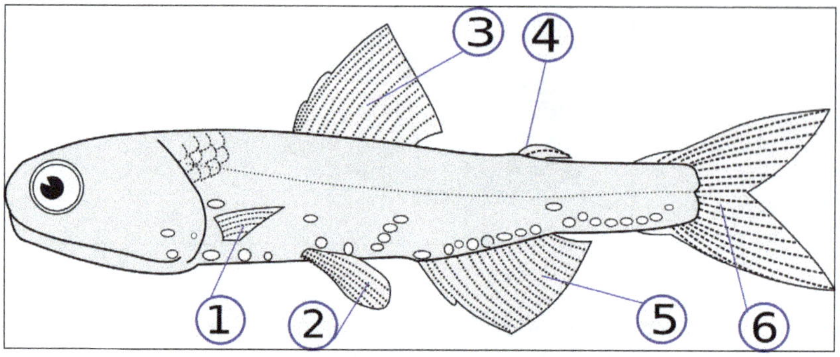

Ray fins on a teleost fish, Hector's lanternfish
(1) pectoral fins (paired), (2) pelvic fins (paired), (3) dorsal fin,
(4) adipose fin, (5) anal fin, (6) caudal (tail) fin.

Types

For every type of fin, there are a number of fish species in which this particular fin has been lost during evolution.

Pectoral Fins

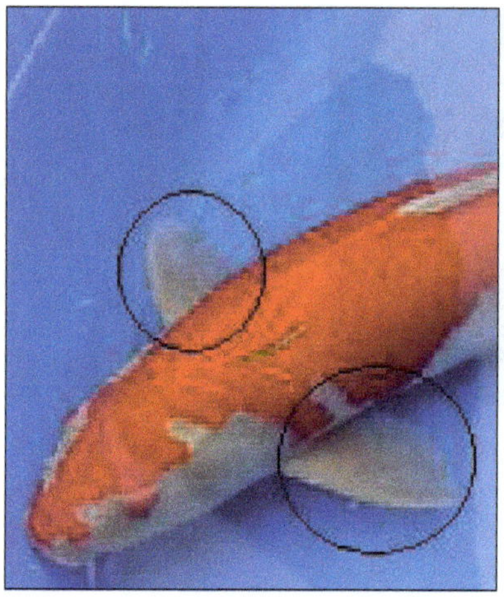

The paired pectoral fins are located on each side, usually just behind the operculum, and are homologous to the forelimbs of tetrapods.

- A peculiar function of pectoral fins, highly developed in some fish, is the creation of the dynamic lifting force that assists some fish, such as sharks, in maintaining depth and also enables the "flight" for flying fish.
- In many fish, the pectoral fins aid in walking, especially in the lobe-like fins of some anglerfish and in the mudskipper.
- Certain rays of the pectoral fins may be adapted into finger-like projections, such as in sea robins and flying gurnards.
    - o The "horns" of manta rays and their relatives are called cephalic fins; this is actually a modification of the anterior portion of the pectoral fin.

Pelvic Fins (Ventral Fins)

The paired pelvic or ventral fins are typically located ventrally below and behind the pectoral fins, although in many fish families they may be positioned in front of the pectoral fins (e.g. cods). They are homologous to the hindlimbs of tetrapods. The pelvic fin assists the fish in going up or down through the water, turning sharply, and stopping quickly.

- In gobies, the pelvic fins are often fused into a single sucker disk. This can be used to attach to objects.
- Pelvic fins can take many positions along the ventral surface of the fish. The ancestral *abdominal* position is seen in (for example) the minnows; the *thoracic* position in sunfish; and the *jugular* position, when the pelvics are anterior to the pectoral fins, as seen in the burbot.

Dorsal Fins

Dorsal fin of a shark.

Dorsal fins are located on the back. A fish can have up to three dorsal fins. The dorsal fins serve to protect the fish against rolling, and assist it in sudden turns and stops.

- In anglerfish, the anterior of the dorsal fin is modified into an illicium and esca, a biological equivalent to a fishing rod and lure.
- The bones that support the dorsal fin are called *Pterygiophore*. There are two to three of them: "proximal", "middle", and "distal". In rock-hard, spinous fins the distal is often fused to the middle, or not present at all.

Dorsal fin of a chub (*Leuciscus cephalus*).

Anal/cloacal Fin

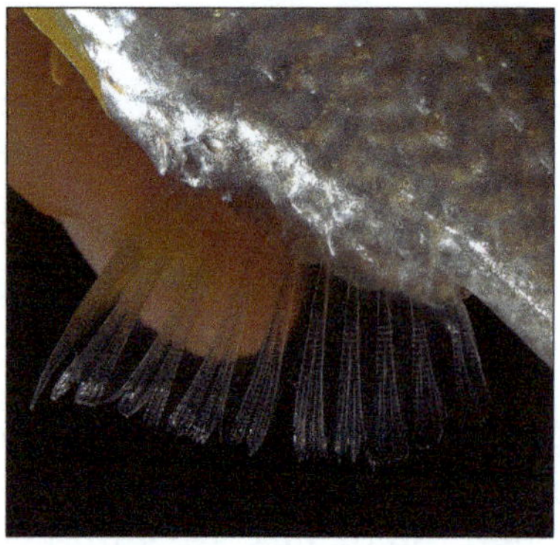

The anal/cloacal fin is located on the ventral surface behind the anus/cloaca. This fin is used to stabilize the fish while swimming.

Adipose Fin

Adipose fin of a trout.

The adipose fin is a soft, fleshy fin found on the back behind the dorsal fin and just forward of the caudal fin. It is absent in many fish families, but found in nine of the 31 euteleostean orders (Percopsiformes, Myctophiformes, Aulopiformes, Stomiiformes, Salmoniformes, Osmeriformes, Characiformes, Siluriformes and Argentiniformes). Famous representatives of these orders are salmon, characids and catfish.

The function of the adipose fin is something of a mystery. It is frequently clipped off to mark hatchery-raised fish, though data from 2005 showed that trout with their adipose fin removed have an 8% higher tailbeat frequency. Additional information released in 2011 has suggested that the fin may be vital for the detection of, and response to, stimuli such as touch, sound and changes in pressure. Canadian researchers identified a neural network in the fin, indicating that it likely has a sensory function, but are still not sure exactly what the consequences of removing it are.

A comparative study in 2013 indicates the adipose fin can develop in two different ways. One is the salmoniform-type way, where the adipose fin develops from the larval-fin fold at the same time and in the same direct manner as the other median fins. The other is the characiform-type way, where the adipose fin develops late after the larval-fin fold has diminished and the other median fins have developed. They claim the existence of the characiform-type of development suggests the adipose fin is not "just a larval fin fold remainder" and is inconsistent with the view that the adipose fin lacks function.

Research published in 2014 indicates that the adipose fin has evolved repeatedly in separate lineages.

Caudal Fin (Tail Fin)

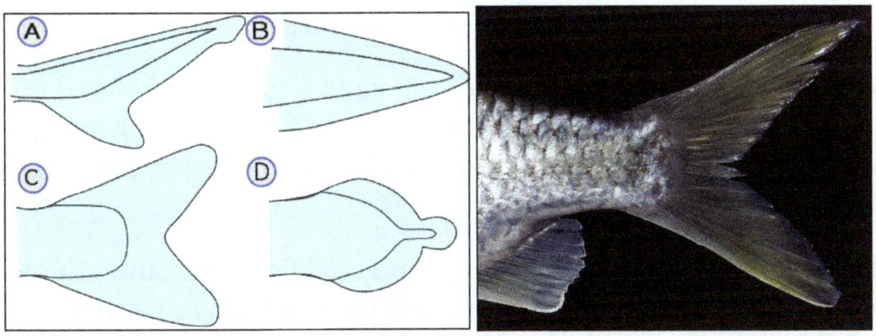

The caudal fin is the tail fin (from the Latin cauda meaning tail), located at the end of the caudal peduncle and is used for propulsion. See body-caudal fin locomotion.

(A) - Heterocercal means the vertebrae extend into the upper lobe of the tail, making it longer (as in sharks).

- Reversed heterocercal means that the vertebrae extend into the lower lobe of the tail, making it longer (as in the Anaspida).

(B) - Protocercal means the vertebrae extend to the tip of the tail and the tail is symmetrical but not expanded (as in amphioxus).

(C) - Homocercal where the fin appears superficially symmetric but in fact the vertebrae extend for a very short distance into the upper lobe of the fin.

(D) - Diphycercal means the vertebrae extend to the tip of the tail and the tail is symmetrical and expanded (as in the bichir, lungfish, lamprey and coelacanth). Most Palaeozoic fishes had a diphycercal heterocercal tail.

Most modern fishes have a homocercal tail. These appear in a variety of shapes, and can appear:

- Rounded.
- Truncated, ending in a more-or-less vertical edge (such as salmon).

- Forked, ending in two prongs.
- Emarginate, ending with a slight inward curve.
- Lunate or shaped like a crescent moon.

Caudal Keel Finlets

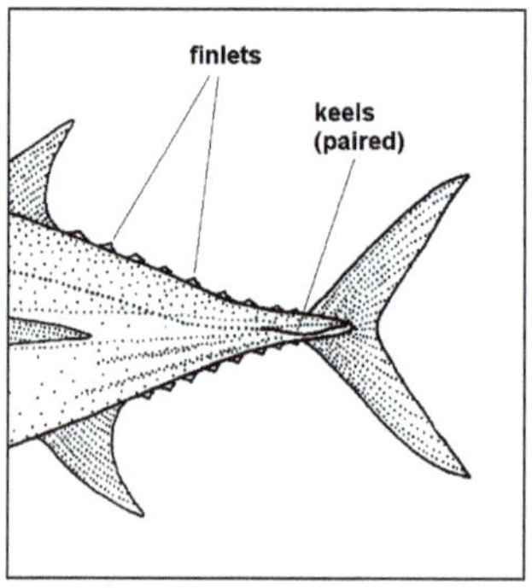

Drawing by Dr Tony Ayling.

Some types of fast-swimming fish have a horizontal caudal keel just forward of the tail fin. Much like the keel of a ship, this is a lateral ridge on the caudal peduncle, usually composed of scutes (see below), that provides stability and support to the caudal fin. There may be a single paired keel, one on each side, or two pairs above and below.

Finlets are small fins, generally behind the dorsal and anal fins (in bichirs, there are only finlets on the dorsal surface and no dorsal fin). In some fish such as tuna or sauries, they are rayless, non-retractable, and found between the last dorsal and/or anal fin and the caudal fin.

Bony Fishes

Skeleton of a ray-finned fish.

Bony fishes form a taxonomic group called Osteichthyes. They have skeletons made of bone, and can be contrasted with cartilaginous fishes which have skeletons made of cartilage. Bony fishes are divided into ray-finned and lobe-finned fish. Most fish are ray-finned, an extremely diverse and abundant group consisting of over 30,000 species. It is the largest class of vertebrates in existence today. In the distant past, lobe-finned fish were abundant. Nowadays they are mainly extinct, with only eight living species. Bony fish have fin spines and rays called lepidotrichia. They typically have swim bladders, which allows the fish to create a neutral balance between sinking and floating without having to use its fins. However, these are absent in many species, and have developed into primitive lungs in the lungfishes. Bony fishes also have an operculum, which helps them breathe without having to use fins to swim.

Lobe-finned

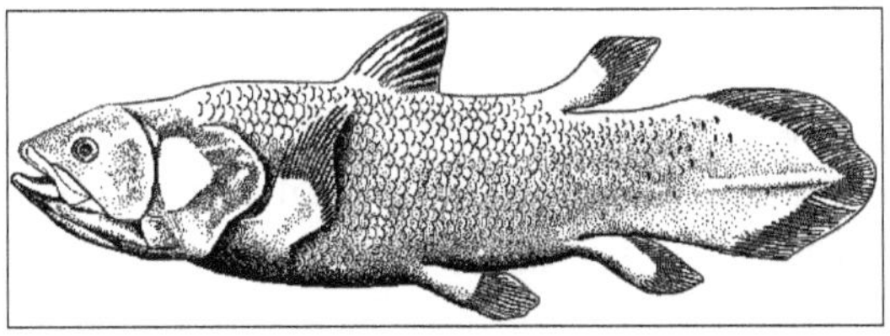

Lobe-finned fishes, like this coelacanth, have fins that are borne on a fleshy, lobelike, scaly stalk extending from the body. Due to the high number of fins it possesses, the coelacanth has high maneuverability and can orient their bodies in almost any direction in the water.

Lobe-finned fishes are a class of bony fishes called Sarcopterygii. They have fleshy, lobed, paired fins, which are joined to the body by a single bone. The fins of lobe-finned fish differ from those of all other fish in that each is borne on a fleshy, lobelike, scaly stalk extending from the body. Pectoral and pelvic fins have articulations resembling those of tetrapod limbs. These fins evolved into legs of the first tetrapod land vertebrates, amphibians. They also possess two dorsal fins with separate bases, as opposed to the single dorsal fin of ray-finned fish.

The coelacanth is another lobe-finned fish which is still extant. It is thought to have evolved into roughly its current form about 408 million years ago, during the early Devonian. Locomotion of the coelacanths is unique to their kind. To move around, coelacanths most commonly take advantage of up or downwellings of the current and drift. They use their paired fins to stabilize their movement through the water. While on the ocean floor their paired fins are not used for any kind of movement. Coelacanths can create thrust for quick starts by using their caudal fins. Due to the high number of fins they possess, coelacanths have high maneuverability and can orient their bodies in almost any direction in the water. They have been seen doing headstands and swimming belly up. It is thought that their rostral organ helps give the coelacanth electroperception, which aids in their movement around obstacles.

Ray-finned

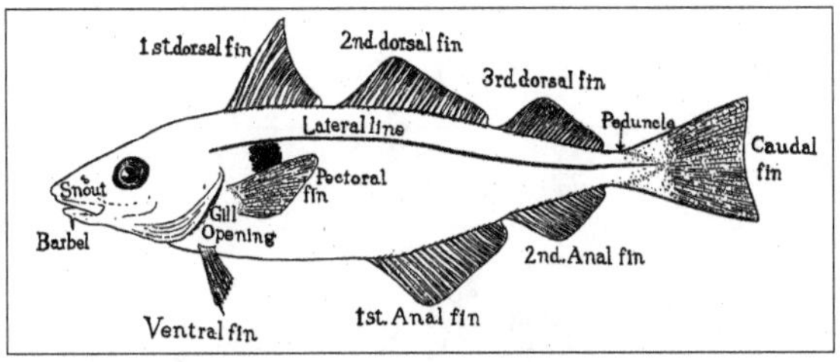

The haddock, a type of cod, is ray-finned. It has three dorsal and two anal fins.

Ray-finned fishes are a class of bony fishes called Actinopterygii. Their fins contain spines or rays. A fin may contain only spiny rays, only soft rays, or a combination of both. If both are present, the spiny rays are always anterior. Spines are generally stiff and sharp. Rays are generally soft, flexible, segmented, and may be branched. This segmentation of rays is the main difference that separates them from spines; spines may be flexible in certain species, but they will never be segmented.

Spines have a variety of uses. In catfish, they are used as a form of defense; many catfish have the ability to lock their spines outwards. Triggerfish also use spines to lock themselves in crevices to prevent them being pulled out.

Lepidotrichia are bony, bilaterally paired, segmented fin rays found in bony fishes. They develop around actinotrichia as part of the dermal exoskeleton. Lepidotrichia are usually composed of bone, but in early osteichthyans such as *Cheirolepis*, there was also dentine and enamel. They are segmented and appear as a series of disks stacked one on top of another. The genetic basis for the formation of the fin rays is thought to be genes coded for the production of certain proteins. It has been suggested that the evolution of the tetrapod limb from lobe-finned fishes is related to the loss of these proteins.

Cartilaginous Fishes

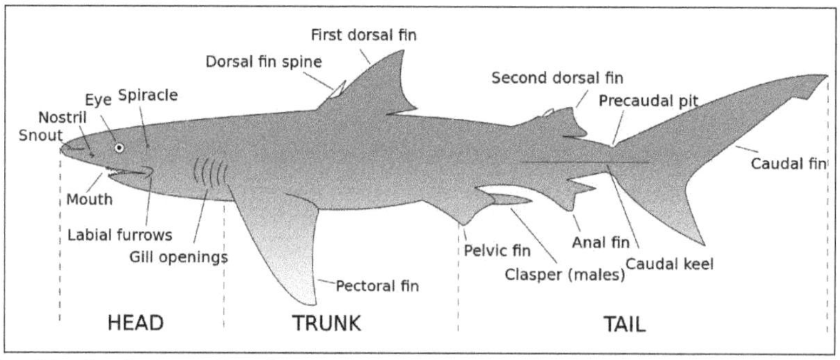

Cartilaginous fishes, like this shark, have fins that are elongated and supported with soft and unsegmented rays named ceratotrichia, filaments of elastic protein resembling the horny keratin in hair and feathers.

Cartilaginous fishes are a class of fishes called Chondrichthyes. They have skeletons made of cartilage rather than bone. The class includes sharks, rays and chimaeras. Shark fin skeletons are elongated and supported with soft and unsegmented rays named ceratotrichia, filaments of elastic protein resembling the horny keratin in hair and feathers. Originally the pectoral and pelvic girdles, which do not contain any dermal elements, did not connect. In later forms, each pair of fins became ventrally connected in the middle when scapulocoracoid and pubioischiadic bars evolved. In rays, the pectoral fins have connected to the head and are very flexible. One of the primary characteristics present in most sharks is the heterocercal tail, which aids in locomotion. Most sharks have eight fins. Sharks can only drift away from objects directly in front of them because their fins do not allow them to move in the tail-first direction.

Caudal fin of a grey reef shark.

As with most fish, the tails of sharks provide thrust, making speed and acceleration dependent on tail shape. Caudal fin shapes vary considerably between shark species, due to their evolution in separate environments. Sharks possess a heterocercal caudal fin in which the dorsal portion is usually noticeably larger than the ventral portion. This is because the shark's vertebral column extends into that dorsal portion, providing a greater surface area for muscle attachment. This allows more efficient locomotion among these negatively buoyant cartilaginous fish. By contrast, most bony fish possess a homocercal caudal fin.

Tiger sharks have a large upper lobe, which allows for slow cruising and sudden bursts of speed. The tiger shark must be able to twist and turn in the water easily when hunting to support its varied diet, whereas the porbeagle shark, which hunts schooling fish such as mackerel and herring, has a large lower lobe to help it keep pace with its fast-swimming prey. Other tail adaptations help sharks catch prey more directly, such as the thresher shark's usage of its powerful, elongated upper lobe to stun fish and squid.

Generating Thrust

Foil shaped fins generate thrust when moved, the lift of the fin sets water or air in motion and pushes the fin in the opposite direction. Aquatic animals get significant thrust by moving fins back and forth in water. Often the tail fin is used, but some aquatic animals generate thrust from pectoral fins.

Moving Fins can Provide Thrust

Fish get thrust moving vertical tail fins from side to side.

Stingrays get thrust from large pectoral fins.

Cavitation occurs when negative pressure causes bubbles (cavities) to form in a liquid, which then promptly and violently collapse. It can cause significant damage and wear. Cavitation damage can occur to the tail fins of powerful swimming marine animals, such as dolphins and tuna. Cavitation is more likely to occur near the surface of the ocean, where the ambient water pressure is relatively low. Even if they have the power to swim faster, dolphins may have to restrict their speed because collapsing cavitation bubbles on their tail are too painful. Cavitation also slows tuna, but for a different reason. Unlike dolphins, these fish do not feel the bubbles, because they have bony fins without nerve endings. Nevertheless, they cannot swim faster because the cavitation bubbles create a vapor film around their fins that limits their speed. Lesions have been found on tuna that are consistent with cavitation damage.

Scombrid fishes (tuna, mackerel and bonito) are particularly high-performance swimmers. Along the margin at the rear of their bodies is a line of small rayless, non-retractable fins, known as finlets. There has been much speculation about the function of these finlets. Research done in 2000 and 2001 by Nauen and Lauder indicated that "the finlets have a hydrodynamic effect on local flow during steady swimming" and that "the most posterior finlet is oriented to redirect flow into the developing tail vortex, which may increase thrust produced by the tail of swimming mackerel".

Fish use multiple fins, so it is possible that a given fin can have a hydrodynamic interaction with another fin. In particular, the fins immediately upstream of the caudal (tail) fin may be proximate fins that can directly affect the flow dynamics at the caudal fin. In 2011, researchers using volumetric imaging techniques were able to generate "the first instantaneous three-dimensional views of wake structures as they are produced by freely swimming fishes". They found that "continuous tail beats resulted in the formation of a linked chain of vortex rings" and that "the dorsal and anal fin wakes are rapidly entrained by the caudal fin wake, approximately within the timeframe of a subsequent tail beat".

Controlling Motion

Once motion has been established, the motion itself can be controlled with the use of other fins.

Specialised Fins are used to Control Motion

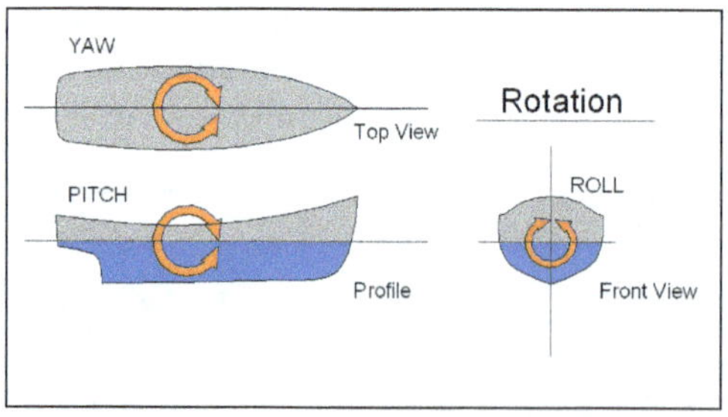

Like boats and airplanes, fish need some control over six degrees of freedom, three translational (heaving, swaying and surging) and three rotational (pitching, yawing and rolling).

Many reef fish have pectoral and pelvic fins optimised for flattened bodies.

The dorsal fin of a white shark contain dermal fibers that work "like riggings that stabilize a ship's mast", and stiffen dynamically as the shark swims faster to control roll and yaw.

The bodies of reef fishes are often shaped differently from open water fishes. Open water fishes are usually built for speed, streamlined like torpedoes to minimise friction as they move through the water. Reef fish operate in the relatively confined spaces and complex underwater landscapes of coral reefs. For this manoeuvrability is more important than straight line speed, so coral reef fish have developed bodies which optimize their ability to dart and change direction. They outwit predators by dodging into fissures in the reef or playing hide and seek around coral heads. The pectoral and pelvic fins of many reef fish, such as butterflyfish, damselfish and angelfish, have evolved so they can act as brakes and allow complex manoeuvres. Many reef fish, such as butterflyfish, damselfish and angelfish, have evolved bodies which are deep and laterally compressed like a pancake, and will fit into fissures in rocks. Their pelvic and pectoral fins are designed differently, so they act together with the flattened body to optimise manoeuvrability. Some fishes, such as puffer fish, filefish and trunkfish, rely on pectoral fins for swimming and hardly use tail fins at all.

Reproduction

This male mosquitofish has a gonopodium, an anal fin which functions as an intromittent organ.

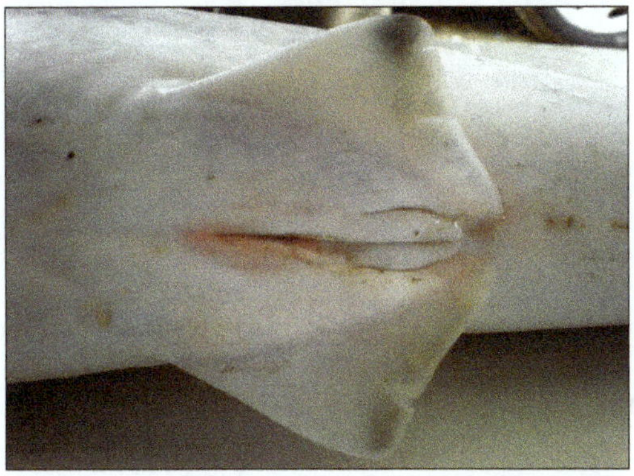

This young male spinner shark has claspers, a modification to the pelvic fins which also function as intromittent organs.

Male cartilaginous fishes (sharks and rays), as well as the males of some live-bearing ray finned fishes, have fins that have been modified to function as intromittent organs, reproductive appendages which allow internal fertilization. In ray finned fish they are called *gonopodia* or *andropodia*, and in cartilaginous fish they are called *claspers*.

Gonopodia are found on the males of some species in the Anablepidae and Poeciliidae families. They are anal fins that have been modified to function as movable intromittent organs and are used to impregnate females with milt during mating. The third, fourth and fifth rays of the male's anal fin are formed into a tube-like structure in which the sperm of the fish is ejected. When ready for mating, the gonopodium becomes erect and points forward towards the female. The male shortly inserts the organ into the sex opening of the female, with hook-like adaptations that allow the fish to grip onto the female to ensure impregnation. If a female remains stationary and her partner contacts her vent with his gonopodium, she is fertilized. The sperm is preserved in the female's oviduct. This allows females to fertilize themselves at any time without further assistance from males. In some species, the gonopodium may be half the total body length. Occasionally the fin is too long to be used, as in the "lyretail" breeds of *Xiphophorus helleri*. Hormone treated females may develop gonopodia. These are useless for breeding.

Similar organs with similar characteristics are found in other fishes, for example the *andropodium* in the *Hemirhamphodon* or in the Goodeidae.

Claspers are found on the males of cartilaginous fishes. They are the posterior part of the pelvic fins that have also been modified to function as intromittent organs, and are used to channel semen into the female's cloaca during copulation. The act of mating in sharks usually includes raising one of the claspers to allow water into a siphon through a specific orifice. The clasper is then inserted into the cloaca, where it opens like an umbrella to anchor its position. The siphon then begins to contract expelling water and sperm.

Other Uses

The Indo-Pacific sailfish has a prominent dorsal fin. Like scombroids and other billfish, they streamline themselves by retracting their dorsal fins into a grove in their body when they swim. The huge dorsal fin, or sail, of the sailfish is kept retracted most of the time. Sailfish raise them if they want to herd a school of small fish, and also after periods of high activity, presumably to cool down.

Frogfish use their pectoral and pelvic fins to walk along the ocean bottom.

The oriental flying gurnard has large pectoral fins which it normally holds against its body, and expands when threatened to scare predators. Despite its name, it is a demersal fish, not a flying fish, and uses its pelvic fins to walk along the bottom of the ocean.

Flying fish achieve sufficient lift to glide above the surface of the water thanks to their enlarged pectoral fins.

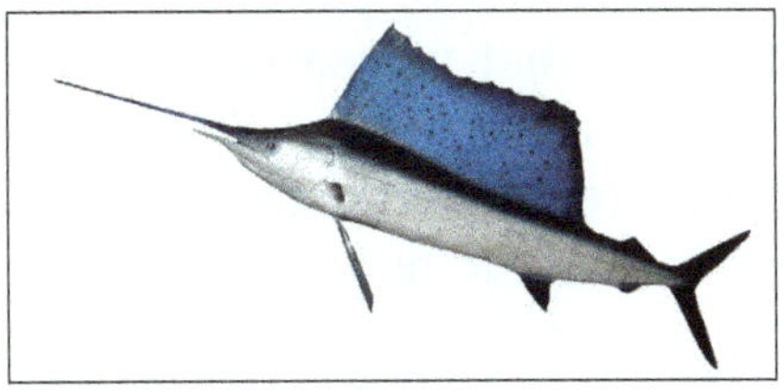

Large retractable dorsal fin of the Indo-Pacific sailfish.

Fins can have an adaptive significance as sexual ornaments. During courtship, the female cichlid, *Pelvicachromis taeniatus*, displays a large and visually arresting purple pelvic fin. "The researchers found that males clearly preferred females with a larger pelvic fin and that pelvic fins grew in a more disproportionate way than other fins on female fish."

The thresher shark uses its caudal fin to stun prey.

Other uses of Fins

The Oriental flying gurnard has large pectoral fins with eye spots which it displays to scare predators.

During courtship, the female cichlid, *Pelvicachromis taeniatus*, displays her visually arresting purple pelvic fin.

Triggerfish squeeze into coral crevices to avoid predators, and lock themselves in place with the first spine of their dorsal fin.

The first spine of the dorsal fin of the anglerfish is modified so it functions like a fishing rod with a lure.

In some Asian countries shark fins are a culinary delicacy.

Evolution

Evolution of Paired Fins

There are two prevailing hypotheses that have been historically debated as models for the evolution of paired fins in fish: the gill arch theory and the lateral fin-fold theory. The former, commonly referred to as the "Gegenbaur hypothesis," was posited in 1870 and proposes that the "paired fins are derived from gill structures". This fell out of popularity in favor of the lateral fin-fold theory, first suggested in 1877, which proposes that paired fins budded from longitudinal, lateral folds along the epidermis just behind the gills. There is weak support for both hypotheses in the fossil record and in embryology. However, recent insights from developmental patterning have prompted reconsideration of both theories in order to better elucidate the origins of paired fins.

Classical Theories

Karl Gegenbaur's concept of the "Archipterygium" was introduced in 1876. It was described as a gill ray, or "joined cartilaginous stem," that extended from the gill arch. Additional rays arose from along the arch and from the central gill ray. Gegenbaur suggested a model of transformative homology – that all vertebrate paired fins and limbs were transformations of the Archipterygium. Based on this theory, paired appendages such as pectoral and pelvic fins would have differentiated from the branchial arches and migrated posteriorly. However, there has been limited support for this hypothesis in the fossil record both morphologically and phylogenically. In addition, there was little to no evidence of an anterior-posterior migration of pelvic fins. Such shortcomings of the gill-arch theory led to its early demise in favor of the lateral fin-fold theory proposed by St. George Jackson Mivart, Francis Balfour, and James Kingsley Thacher.

The lateral fin-fold theory hypothesized that paired fins developed from lateral folds along the body wall of the fish. Just as segmentation and budding of the median fin fold gave rise to the median fins, a similar mechanism of fin bud segmentation and elongation from a lateral fin fold was proposed to have given rise to the paired pectoral and pelvic fins. However, there was little evidence of a lateral fold-to-fin transition in the fossil record. In addition, it was later demonstrated phylogenically that pectoral and pelvic fins arise from distinct evolutionary and mechanistic origins.

Evolutionary Developmental Biology

Recent studies in the ontogeny and evolution of paired appendages have compared finless vertebrates – such as lampreys – with chondricthyes, the most basal living vertebrate with paired fins. In 2006, researchers found that the same genetic programming involved in the segmentation and development of median fins was found in the development of paired appendages in catsharks. Although these findings do not directly support the lateral fin-fold hypothesis, the original concept of a shared median-paired fin evolutionary developmental mechanism remains relevant.

We find a similar renovation of an old theory in the developmental programming of chondricthyan gill arches and paired appendages. In 2009, researchers at the University of Chicago demonstrated that there are shared molecular patterning mechanisms in the early development of the chondricthyan gill arch and paired fins. Findings such as these have prompted reconsideration of the once-debunked gill-arch theory.

From Fins to Limbs

Fish are the ancestors of all mammals, reptiles, birds and amphibians. In particular, terrestrial tetrapods (four-legged animals) evolved from fish and made their first forays onto land 400 million years ago. They used paired pectoral and pelvic fins for locomotion. The pectoral fins developed into forelegs (arms in the case of humans) and the pelvic fins developed into hind legs. Much of the genetic machinery that builds a walking limb in a tetrapod is already present in the swimming fin of a fish.

Aristotle recognised the distinction between analogous and homologous structures, and made the following prophetic comparison: *"Birds in a way resemble fishes. For birds have their wings in the upper part of their bodies and fishes have two fins in the front part of their bodies. Birds have feet on their underpart and most fishes have a second pair of fins in their under-part and near their front fins."*

– Aristotle, De incessu animalium

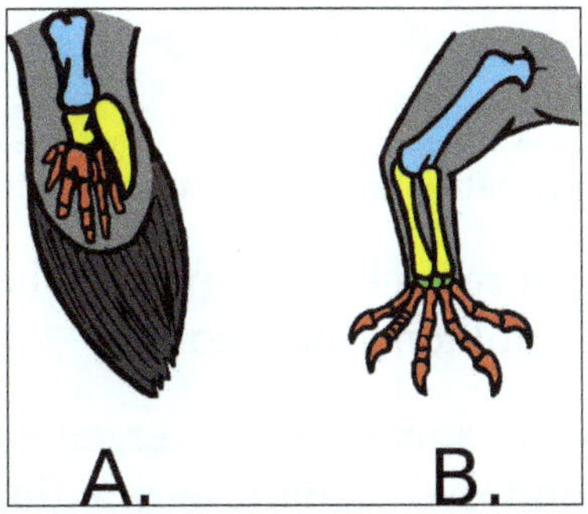

Comparison between A) the swimming fin of a lobe-finned fish and B) the walking leg of a tetrapod. Bones considered to correspond with each other have the same color.

In a parallel but independent evolution, the ancient reptile *Ichthyosaurus communis* developed fins (or flippers) very similar to fish (or dolphins).

In 2011, researchers at Monash University in Australia used primitive but still living lungfish "to trace the evolution of pelvic fin muscles to find out how the load-bearing hind limbs of the tetrapods evolved." Further research at the University of Chicago found bottom-walking lungfishes had already evolved characteristics of the walking gaits of terrestrial tetrapods.

In a classic example of convergent evolution, the pectoral limbs of pterosaurs, birds and bats further evolved along independent paths into flying wings. Even with flying wings there are many similarities with walking legs, and core aspects of the genetic blueprint of the pectoral fin have been retained.

The first mammals appeared during the Permian period (between 298.9 and 252.17 million years ago). Several groups of these mammals started returning to the sea, including the cetaceans (whales, dolphins and porpoises). Recent DNA analysis suggests that cetaceans evolved from within the even-toed ungulates, and that they share a common ancestor with the hippopotamus. About 23 million years ago another group of bearlike land mammals started returning to the sea. These were the seals. What had become walking limbs in cetaceans and seals evolved independently into new forms of swimming fins. The forelimbs became flippers, while the hindlimbs were either lost (cetaceans) or also modified into flipper (pinnipeds). In cetaceans, the tail gained two fins at the end, called a fluke. Fish tails are usually vertical and move from side to side. Cetacean flukes are horizontal and move up and down, because cetacean spines bend the same way as in other mammals.

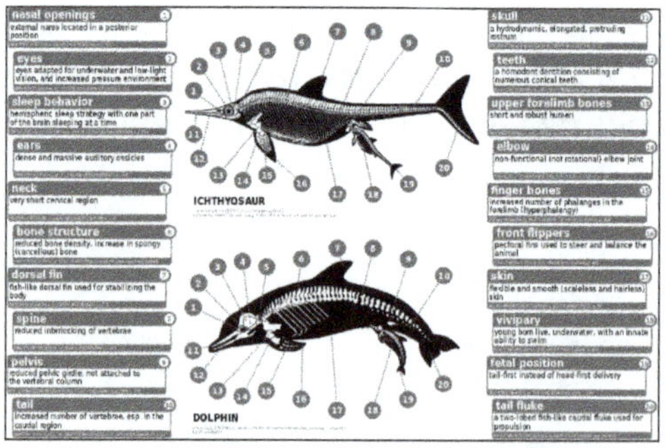

Similar adaptations for fully aquatic lifestyle are found both in dolphins and ichthyosaurs.

Ichthyosaurs are ancient reptiles that resembled dolphins. They first appeared about 245 million years ago and disappeared about 90 million years ago.

"This sea-going reptile with terrestrial ancestors converged so strongly on fishes that it actually evolved a dorsal fin and tail fin for improved aquatic locomotion. These structures are all the more remarkable because they evolved from nothing — the ancestral terrestrial reptile had no hump on its back or blade on its tail to serve as a precursor."

The biologist Stephen Jay Gould said the ichthyosaur was his favorite example of convergent evolution.

Fins or flippers of varying forms and at varying locations (limbs, body, tail) have also evolved in a number of other tetrapod groups, including diving birds such as penguins (modified from wings), sea turtles (forelimbs modified into flippers), mosasaurs (limbs modified into flippers), and sea snakes (vertically expanded, flattened tail fin).

Robotic Fins

In the 1990s, the CIA built a robotic catfish called *Charlie*, designed to collect underwater intelligence undetected.

The use of fins for the propulsion of aquatic animals can be remarkably effective. It has been calculated that some fish can achieve a propulsive efficiency greater than 90%. Fish can accelerate and maneuver much more effectively than boats or submarine, and produce less water disturbance and noise. This has led to biomimetic studies of underwater robots which attempt to emulate the locomotion of aquatic animals. An example is the Robot Tuna built by the Institute of Field Robotics, to analyze and mathematically model thunniform motion. In 2005, the Sea Life London Aquarium displayed three robotic fish created by the computer science department at the University of Essex. The fish were designed to be autonomous, swimming around and avoiding obstacles like real fish. Their creator claimed that he was trying to combine "the speed of tuna, acceleration of a pike, and the navigating skills of an eel."

The *AquaPenguin*, developed by Festo of Germany, copies the streamlined shape and propulsion by front flippers of penguins. Festo also developed *AquaRay*, *AquaJelly* and *AiraCuda*, respectively emulating the locomotion of manta rays, jellyfish and barracuda.

In 2004, Hugh Herr at MIT prototyped a biomechatronic robotic fish with a living actuator by surgically transplanting muscles from frog legs to the robot and then making the robot swim by pulsing the muscle fibers with electricity.

Robotic fish offer some research advantages, such as the ability to examine an individual part of a fish design in isolation from the rest of the fish. However, this risks oversimplifying the biology so key aspects of the animal design are overlooked. Robotic fish also allow researchers to vary a single parameter, such as flexibility or a specific motion control. Researchers can directly measure forces, which is not easy to do in live fish. "Robotic devices also facilitate three-dimensional kinematic studies and correlated hydrodynamic analyses, as the location of the locomotor surface can be known accurately. And, individual components of a natural motion (such as outstroke vs. instroke of a flapping appendage) can be programmed separately, which is certainly difficult to achieve when working with a live animal."

Fish Locomotion

Fish propel themselves through water using many different mechanisms.

Fish locomotion is the variety of types of animal locomotion used by fish, principally by swimming. This however is achieved in different groups of fish by a variety of mechanisms of propulsion in water, most often by wavelike movements of the fish's body and tail, and in various specialised fish by movements of the fins. The major forms of locomotion in fish are anguilliform, in which a wave passes evenly along a long slender body; sub-carangiform, in which the wave increases quickly in amplitude towards the tail; carangiform, in which the wave is concentrated near the tail, which oscillates rapidly; thunniform, rapid swimming with a large powerful crescent-shaped tail; and ostraciiform, with almost no oscillation except of the tail fin. More specialised fish include movement by pectoral fins with a mainly stiff body, as in the sunfish; and movement by propagating a wave along the long fins with a motionless body in fish with electric organs such as the knifefish.

In addition, some fish can variously "walk", i.e., move over land, burrow in mud, and glide through the air.

Swimming

Fish swim by exerting force against the surrounding water. There are exceptions, but this is normally achieved by the fish contracting muscles on either side of its body in order to generate waves of flexion that travel the length of the body from nose to tail, generally getting larger as they go along. The vector forces exerted on the water by such motion cancel out laterally, but generate a net force backwards which in turn pushes the fish forward through the water. Most fishes generate thrust using lateral movements of their body and caudal fin, but many other species move mainly using their median and paired fins. The latter group swim slowly, but can turn rapidly, as is needed when living in coral reefs for example. But they can't swim as fast as fish using their bodies and caudal fins.

Body/caudal Fin Propulsion

There are five groups that differ in the fraction of their body that is displaced laterally:

Anguilliform

Eels propagate a more or less constant-sized flexion wave along their slender bodies.

In the anguilliform group, containing some long, slender fish such as eels, there is little increase in the amplitude of the flexion wave as it passes along the body.

Sub-carangiform

The sub-carangiform group has a more marked increase in wave amplitude along the body with the vast majority of the work being done by the rear half of the fish. In general, the fish body is stiffer, making for higher speed but reduced maneuverability. Trout use sub-carangiform locomotion.

Carangiform

The carangiform group, named for the Carangidae, are stiffer and faster-moving than the previous groups. The vast majority of movement is concentrated in the very rear of the body and tail. Carangiform swimmers generally have rapidly oscillating tails.

Thunniform

Tunas such as the bluefin swim fast with their large crescent-shaped tails.

The thunniform group contains high-speed long-distance swimmers, and is a unique trait (an autapomorphy) of the tunas. Here, virtually all the sideways movement is in the tail and the region connecting the main body to the tail (the peduncle). The tail itself tends to be large and crescent shaped.

Ostraciiform

The ostraciiform group have no appreciable body wave when they employ caudal locomotion. Only the tail fin itself oscillates (often very rapidly) to create thrust. This group includes Ostraciidae.

Median/paired Fin Propulsion

Boxfish use median-paired fin swimming, as they are not well streamlined, and use primarily their pectoral fins to produce thrust.

Not all fish fit comfortably in the above groups. Ocean sunfish, for example, have a completely different system, the tetraodontiform mode, and many small fish use their pectoral fins for swimming as well as for steering and dynamic lift. Fish with electric organs, such as those in the knifefish (Gymnotiformes), swim by undulating their very long fins while keeping the body still, presumably so as not to disturb the electric field that they generate.

Many fish swim using combined behavior of their two pectoral fins or both their anal and dorsal fins. Different types of Median paired fin propulsion can be achieved by preferentially using one fin pair over the other, and include rajiform, diodontiform, amiiform, gymnotiform and balistiform modes.

Rajiform

Rajiform locomotion is characteristic of rays, skates, and mantas when thrust is produced by vertical undulations along large, well developed pectoral fins.

Diodontiform

Porcupine fish (here, *Diodon nicthemerus*) swim by undulating their pectoral fins.

Diodontiform locomotion propels the fish propagating undulations along large pectoral fins, as seen in the porcupinefish (Diodontidae).

Amiiform

Amiiform locomotion consists of undulations of a long dorsal fin while the body axis is held straight and stable, as seen in the bowfin.

Gymnotiform

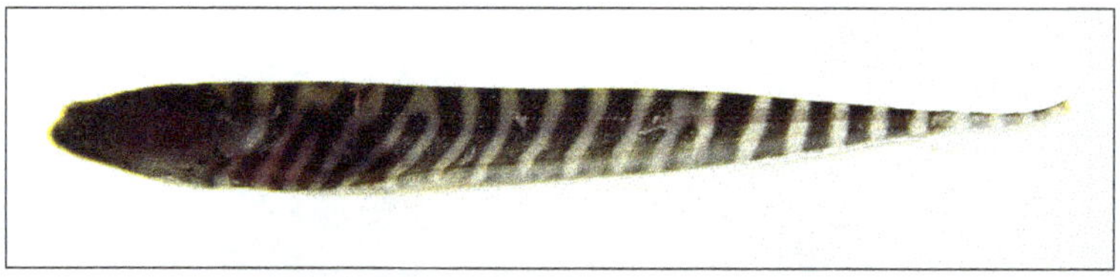

Gymnotus maintains a straight back while swimming to avoid disturbing its electric sense.

Gymnotiform locomotion consists of undulations of a long anal fin, essentially upside down amiiform, seen in the knifefish (Gymnotiformes).

Balistiform

In balistiform locomotion, both anal and dorsal fins undulate, as seen in the Zeidae.

Oscillatory

Oscillation is viewed as pectoral-fin-based swimming and is best known as mobuliform locomotion. The motion can be described as the production of less than half a wave on the fin, similar to a bird wing flapping. Pelagic stingrays, such as the manta, cownose, eagle and bat rays use oscillatory locomotion.

Tetraodontiform

In tetraodontiform locomotion, the dorsal and anal fins are flapped as a unit, either in phase or exactly opposing one another, as seen in the Tetraodontiformes (boxfishes and pufferfishes). The ocean sunfish displays an extreme example of this mode.

Labriform

In labriform locomotion, seen in the wrasses (Labriformes), oscillatory movements of pectoral fins are either drag based or lift based. Propulsion is generated either as a reaction to drag produced by dragging the fins through the water in a rowing motion, or via lift mechanisms.

Dynamic Lift

Sharks are denser than water, and must swim continually, using dynamic lift from their pectoral fins.

Bone and muscle tissues of fish are denser than water. To maintain depth fish such as sharks, but also some bony fish, increase buoyancy by means of a gas bladder or by storing oils or lipids. Fish without these features use dynamic lift instead. It is done using their pectoral fins in a manner similar to the use of wings by airplanes and birds. As these fish swim, their pectoral fins are positioned to create lift which allows the fish to maintain a certain depth. The two major drawbacks of this method are that these fish must stay moving to stay afloat and that they are incapable of swimming backwards or hovering.

Hydrodynamics

Similarly to the aerodynamics of flight, powered swimming requires animals to overcome drag by producing thrust. Unlike flying, however, swimming animals often do not need to supply much vertical force because the effect of buoyancy can counter the downward pull of gravity, allowing these animals to float without much effort. While there is great diversity in fish locomotion, swimming behavior can be classified into two distinct "modes" based on the body structures involved in thrust production, Median-Paired Fin (MPF) and Body-Caudal Fin (BCF). Within each of these classifications, there are numerous specifications along a spectrum of behaviours from purely undulatory to entirely oscillatory. In undulatory swimming modes, thrust is produced by wave-like movements of the propulsive structure (usually a fin or the whole body). Oscillatory modes, on the other hand, are characterized by thrust produced by swiveling of the propulsive structure on an attachment point without any wave-like motion.

Body-caudal Fin

Most fish swim by generating undulatory waves that propagate down the body through the caudal fin. This form of undulatory locomotion is termed Body-Caudal Fin (BCF) swimming on the basis of the body structures used; it includes anguilliform, sub-carangiform, carangiform, and thunniform locomotory modes, as well as the oscillatory ostraciiform mode.

Adaptation

Similar to adaptation in avian flight, swimming behaviors in fish can be thought of as a balance of stability and maneuverability. Because BCF swimming relies on more caudal body structures that can direct powerful thrust only rearwards, this form of locomotion is particularly effective for accelerating quickly and cruising continuously. BCF swimming is, therefore, inherently stable and is often seen in fish with large migration patterns that must maximize efficiency over long periods. Propulsive forces in MPF swimming, on the other hand, are generated by multiple fins located on either side of the body that can be coordinated to execute elaborate turns. As a result, MPF swimming is well adapted for high maneuverability and is often seen in smaller fish that require elaborate escape patterns.

The habitats occupied by fishes are often related to their swimming capabilities. On coral reefs, the faster-swimming fish species typically live in wave-swept habitats subject to fast water flow speeds, while the slower fishes live in sheltered habitats with low levels of water movement.

Fish do not rely exclusively on one locomotor mode, but are rather locomotor generalists, choosing among and combining behaviors from many available behavioral techniques. At slower speeds, predominantly BCF swimmers often incorporate movement of their pectoral, anal, and dorsal fins as an additional stabilizing mechanism at slower speeds, but hold them close to their body at high speeds to improve streamlining and reducing drag. Zebrafish have even been observed to alter their locomotor behavior in response to changing hydrodynamic influences throughout growth and maturation.

In addition to adapting locomotor behavior, controlling buoyancy effects is critical for aquatic survival since aquatic ecosystems vary greatly by depth. Fish generally control their depth by regulating the amount of gas in specialized organs that are much like balloons. By changing the amount of gas in these swim bladders, fish actively control their density. If they increase the amount of air in their swim bladder, their overall density will become less than the surrounding water, and increased upward buoyancy pressures will cause the fish to rise until they reach a depth at which they are again at equilibrium with the surrounding water.

Flight

The transition of predominantly swimming locomotion directly to flight has evolved in a single family of marine fish, the Exocoetidae. Flying fish are not true fliers in the sense that they do not execute powered flight. Instead, these species glide directly over the surface of the ocean water without ever flapping their "wings." Flying fish have evolved abnormally large pectoral fins that act as airfoils and provide lift when the fish launches itself out of the water. Additional forward thrust and steering forces are created by dipping the hypocaudal (i.e. bottom) lobe of their caudal fin into the water and vibrating it very quickly, in contrast to diving birds in which these forces are produced by the same locomotor module used for propulsion. Of the 64 extant species of flying fish, only two distinct body plans exist, each of which optimizes two different behaviors.

Flying fish gain sufficient lift to glide above the water thanks to their enlarged pectoral fins.

Tradeoffs

While most fish have caudal fins with evenly sized lobes (i.e. homocaudal), flying fish have an enlarged ventral lobe (i.e. hypocaudal) which facilitates dipping only a portion of the tail back onto the water for additional thrust production and steering.

Because flying fish are primarily aquatic animals, their body density must be close to that of water for buoyancy stability. This primary requirement for swimming, however, means that flying fish are heavier (have a larger mass) than other habitual fliers, resulting in higher wing loading and lift to drag ratios for flying fish compared to a comparably sized bird. Differences in wing area, wing span, wing loading, and aspect ratio have been used to classify flying fish into two distinct classifications based on these different aerodynamic designs.

Biplane Body Plan

In the biplane or *Cypselurus* body plan, both the pectoral and pelvic fins are enlarged to provide lift during flight. These fish also tend to have "flatter" bodies which increase the total lift producing area thus allowing them to "hang" in the air better than more streamlined shapes. As a result of this high lift production, these fish are excellent gliders and are well adapted for maximizing flight distance and duration.

Comparatively, *Cypselurus* flying fish have lower wing loading and smaller aspect ratios (i.e. broader wings) than their *Exocoetus* monoplane counterparts, which contributes to their ability to fly for longer distances than fish with this alternative body plan. Flying fish with the biplane design take advantage of their high lift production abilities when launching from the water by utilizing a "taxiing glide" in which the hypocaudal lobe remains in the water to generate thrust even after the trunk clears the water's surface and the wings are opened with a small angle of attack for lift generation.

Monoplane Body Plan

In the monoplane body plan of *Exocoetus*, only the pectoral fins are abnormally large, while the pelvic fins are small.

In the *Exocoetus* or monoplane body plan, only the pectoral fins are enlarged to provide lift. Fish with this body plan tend to have a more streamlined body, higher aspect ratios (long, narrow wings), and higher wing loading than fish with the biplane body plan, making these fish well adapted for higher flying speeds. Flying fish with a monoplane body plan demonstrate different launching behaviors from their biplane counterparts. Instead of extending their duration of thrust production, monoplane fish launch from the water at high speeds at a large angle of attack (sometimes up to 45 degrees). In this way, monoplane fish are taking advantage of their adaptation for high flight speed, while fish with biplane designs exploit their lift production abilities during takeoff.

Walking

A "walking fish" is a fish that is able to travel over land for extended periods of time. Some other cases of nonstandard fish locomotion include fish "walking" along the sea floor, such as the handfish or frogfish.

Most commonly, walking fish are amphibious fish. Able to spend longer times out of water, these fish may use a number of means of locomotion, including springing, snake-like lateral undulation, and tripod-like walking. The mudskippers are probably the best land-adapted of contemporary fish and are able to spend days moving about out of water and can even climb mangroves, although to only modest heights. The Climbing gourami is often specifically referred to as a "walking fish", although it does not actually "walk", but rather moves in a jerky way by supporting itself on the extended edges of its gill plates and pushing itself by its fins and tail. Some reports indicate that it can also climb trees.

There are a number of fish that are less adept at actual walking, such as the walking catfish. Despite being known for "walking on land", this fish usually wriggles and may use its pectoral fins to aid in its movement. Walking Catfish have a respiratory system that allows them to live out of water for several days. Some are invasive species. A notorious case in the United States is the Northern snakehead. Polypterids have rudimentary lungs and can also move about on land, though rather clumsily. The Mangrove rivulus can survive for months out of water and can move to places like hollow logs.

Ogcocephalus parvus.

There are some species of fish that can "walk" along the sea floor but not on land; one such animal is the flying gurnard (it does not actually fly, and should not be confused with flying fish). The batfishes of the Ogcocephalidae family are also capable of walking along the sea floor. Bathypterois grallator, also known as a "tripodfish", stands on its three fins on the bottom of the ocean and hunts for food. The African lungfish (*P. annectens*) can use its fins to *"walk"* along the bottom of its tank in a manner similar to the way amphibians and land vertebrates use their limbs on land.

Burrowing

Many fishes, particularly eel-shaped fishes such as true eels, moray eels, and spiny eels, are capable of burrowing through sand or mud. Ophichthids, the snake eels, are capable of burrowing either forwards or backwards.

Swim Bladder

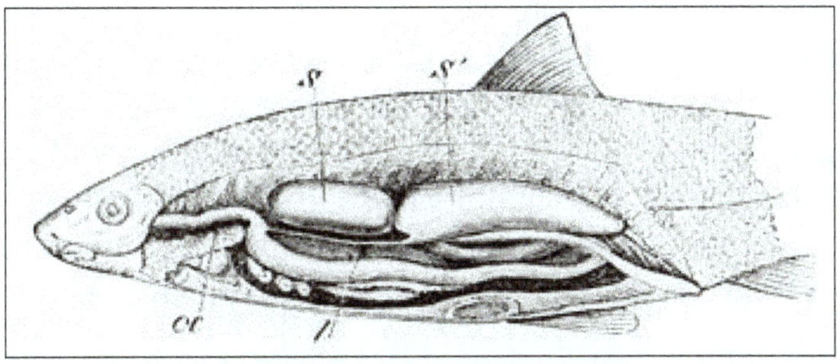

Internal positioning of the swim bladder of a bleak S: anterior, S': posterior portion of the air bladder œ: œsophagus; l: air passage of the air bladder.

The swim bladder, gas bladder, fish maw or air bladder is an internal gas-filled organ that contributes to the ability of many bony fish (but not cartilaginous fish) to control their buoyancy, and thus to stay at their current water depth without having to waste energy in swimming. Also, the dorsal position of the swim bladder means the center of mass is below the center of volume, allowing it to act as a stabilizing agent. Additionally, the swim bladder functions as a resonating chamber, to produce or receive sound.

The swim bladder is evolutionarily homologous to the lungs. Charles Darwin remarked upon this in *On the Origin of Species*. Darwin reasoned that the lung in air-breathing vertebrates had derived from a more primitive swim bladder, but scientists now believe that the swim bladder derived from a more primitive lung.

In the embryonic stages some species, such as redlip blenny, have lost the swim bladder again, mostly bottom dwellers like the weather fish. Other fish like the Opah and the Pomfret use their pectoral fins to swim and balance the weight of the head to keep a horizontal position. The normally bottom dwelling sea robin can use their pectoral fins to produce lift while swimming.

The gas/tissue interface at the swim bladder produces a strong reflection of sound, which is used in sonar equipment to find fish.

The cartilaginous fish (e.g. sharks and rays) do not have swim bladders. Some of them can control their depth only by swimming (using dynamic lift); others store fats or oils with density less than that of seawater to produce a neutral or near neutral buoyancy, which does not change with depth.

Structure and Function

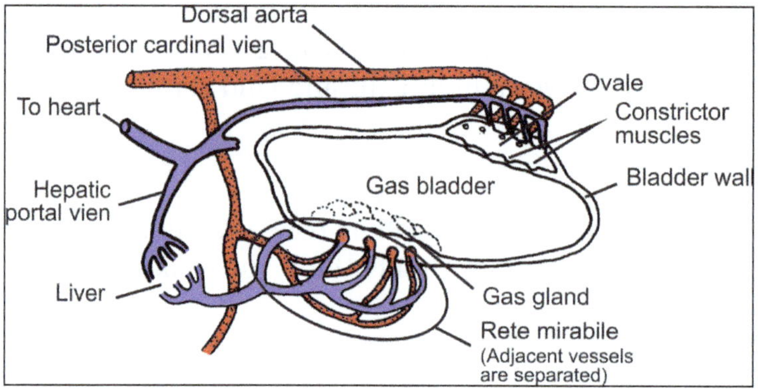

Swim bladder from a bony (teleost) fish.

The swim bladder normally consists of two gas-filled sacs located in the dorsal portion of the fish, although in a few primitive species, there is only a single sac. It has flexible walls that contract or expand according to the ambient pressure. The walls of the bladder contain very few blood vessels and are lined with guanine crystals, which make them impermeable to gases. By adjusting the gas pressurising organ using the gas gland or oval window the fish can obtain neutral buoyancy and ascend and descend to a large range of depths. Due to the dorsal position it gives the fish lateral stability.

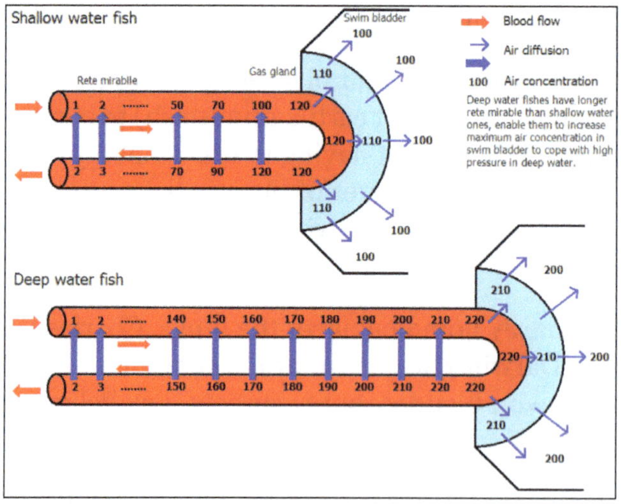

How gas is pumped into the swim bladder using counter-current exchange.

In physostomous swim bladders, a connection is retained between the swim bladder and the gut, the pneumatic duct, allowing the fish to fill up the swim bladder by "gulping" air. Excess gas can be removed in a similar manner.

In more derived varieties of fish (the physoclisti) the connection to the digestive tract is lost. In early life stages, these fish must rise to the surface to fill up their swim bladders; in later stages, the pneumatic duct disappears, and the gas gland has to introduce gas (usually oxygen) to the bladder to increase its volume and thus increase buoyancy. In order to introduce gas into the bladder, the gas gland excretes lactic acid and produces carbon dioxide. The resulting acidity causes the hemoglobin of the blood to lose its oxygen (Root effect) which then diffuses partly into the swim bladder. The blood flowing back to the body first enters a rete mirabile where virtually all the excess carbon dioxide and oxygen produced in the gas gland diffuses back to the arteries supplying the gas gland. Thus a very high gas pressure of oxygen can be obtained, which can even account for the presence of gas in the swim bladders of deep sea fish like the eel, requiring a pressure of hundreds of bars. Elsewhere, at a similar structure known as the *oval window*, the bladder is in contact with blood and the oxygen can diffuse back out again. Together with oxygen, other gases are salted out in the swim bladder which accounts for the high pressures of other gases as well.

The combination of gases in the bladder varies. In shallow water fish, the ratios closely approximate that of the atmosphere, while deep sea fish tend to have higher percentages of oxygen. For instance, the eel *Synaphobranchus* has been observed to have 75.1% oxygen, 20.5% nitrogen, 3.1% carbon dioxide, and 0.4% argon in its swim bladder.

Physoclist swim bladders have one important disadvantage: they prohibit fast rising, as the bladder would burst. Physostomes can "burp" out gas, though this complicates the process of re-submergence.

The swim bladder in some species, mainly fresh water fishes (Common carp, catfish, bowfin) is interconnected with the inner ear of the fish. They are connected by four bones called the Weberian ossicles from the Weberian apparatus. These bones can carry the vibrations to the Saccule and the Lagena (anatomy). They are suited for detecting sound and vibrations due to its low density in comparison to the density of the fish's body tissues. This increases the ability of sound detection. The swim bladder can radiate the pressure of sound which help increase its sensitivity and expand its hearing. In some deep sea fishes like the *Antimora*, the swim bladder maybe also connected to the Macula of saccule in order for the inner ear to receive a sensation from the sound pressure. In red-bellied piranha, the swimbladder may play an important role in sound production as a resonator. The sounds created by piranhas are generated through rapid contractions of the sonic muscles and is associated with the swimbladder.

Evolution

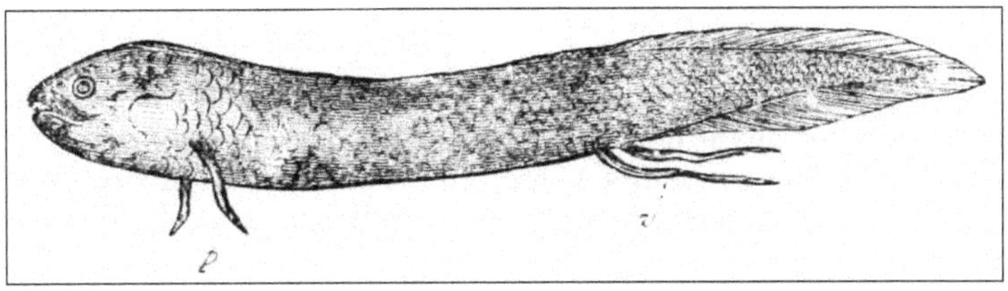

The West African lungfish possesses a lung homologous to swim bladders.

The illustration of the swim bladder in fishes ... shows us clearly the highly important fact that an organ originally constructed for one purpose, namely, flotation, may be converted into one for a widely different purpose, namely, respiration. The swim bladder has, also, been worked in as an accessory to the auditory organs of certain fishes. All physiologists admit that the swimbladder is homologous, or "ideally similar" in position and structure with the lungs of the higher vertebrate animals: hence there is no reason to doubt that the swim bladder has actually been converted into lungs, or an organ used exclusively for respiration. According to this view it may be inferred that all vertebrate animals with true lungs are descended by ordinary generation from an ancient and unknown prototype, which was furnished with a floating apparatus or swim bladder.

Charles Darwin, 1859

Swim bladders are evolutionarily closely related (i.e., homologous) to lungs. Traditional wisdom has long held that the first lungs, simple sacs connected to the gut that allowed the organism to gulp air under oxygen-poor conditions, evolved into the lungs of today's terrestrial vertebrates and some fish (e.g., lungfish, gar, and bichir) and into the swim bladders of the ray-finned fish. In 1997, Farmer proposed that lungs evolved to supply the heart with oxygen. In fish, blood circulates from the gills to the skeletal muscle, and only then to the heart. During intense exercise, the oxygen in the blood gets used by the skeletal muscle before the blood reaches the heart. Primitive lungs gave an advantage by supplying the heart with oxygenated blood via the cardiac shunt. This theory is robustly supported by the fossil record, the ecology of extant air-breathing fishes, and the physiology of extant fishes. In embryonal development, both lung and swim bladder originate as an outpocketing from the gut; in the case of swim bladders, this connection to the gut continues to exist as the pneumatic duct in the more "primitive" ray-finned fish, and is lost in some of the more derived teleost orders. There are no animals which have both lungs and a swim bladder.

The cartilaginous fish (e.g., sharks and rays) split from the other fishes about 420 million years ago, and lack both lungs and swim bladders, suggesting that these structures evolved after that split. Correspondingly, these fish also have both heterocercal and pectoral fins which provide the necessary lift needed due to the lack of swim bladders. Teleost fish with swim bladders have neutral buoyancy, and have no need for this lift.

Deep Scattering Layer

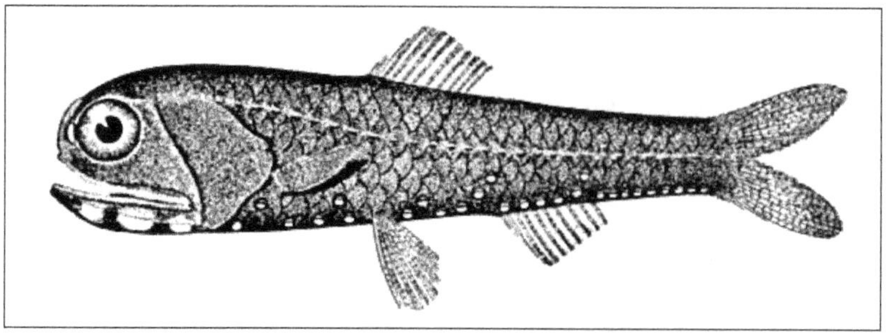

Most mesopelagic fishes are small filter feeders which ascend at night using their swimbladders to feed in the nutrient rich waters of the epipelagic zone. During the day, they return to the dark, cold, oxygen deficient waters of the mesopelagic where they are relatively safe from predators. Lanternfish account for as much as 65 percent of all deep sea fish biomass and are largely responsible for the deep scattering layer of the world's oceans.

Sonar operators, using the newly developed sonar technology during World War II, were puzzled by what appeared to be a false sea floor 300–500 metres deep at day, and less deep at night. This turned out to be due to millions of marine organisms, most particularly small mesopelagic fish, with swimbladders that reflected the sonar. These organisms migrate up into shallower water at dusk to feed on plankton. The layer is deeper when the moon is out, and can become shallower when clouds obscure the moon.

Most mesopelagic fish make daily vertical migrations, moving at night into the epipelagic zone, often following similar migrations of zooplankton, and returning to the depths for safety during the day. These vertical migrations often occur over large vertical distances, and are undertaken with the assistance of a swim bladder. The swim bladder is inflated when the fish wants to move up, and, given the high pressures in the mesoplegic zone, this requires significant energy. As the fish ascends, the pressure in the swimbladder must adjust to prevent it from bursting. When the fish wants to return to the depths, the swimbladder is deflated. Some mesopelagic fishes make daily migrations through the thermocline, where the temperature changes between 10 and 20°C, thus displaying considerable tolerance for temperature change.

Sampling via deep trawling indicates that lanternfish account for as much as 65% of all deep sea fish biomass. Indeed, lanternfish are among the most widely distributed, populous, and diverse of all vertebrates, playing an important ecological role as prey for larger organisms. The estimated global biomass of lanternfish is 550–660 million metric tonnes, several times the entire world fisheries catch. Lanternfish also account for much of the biomass responsible for the deep scattering layer of the world's oceans. Sonar reflects off the millions of lanternfish swim bladders, giving the appearance of a false bottom.

Human Uses

In some Asian cultures, the swim bladders of certain large fishes are considered a food delicacy. In China they are known as *fish maw* and are served in soups or stews.

The vanity price of a vanishing kind of maw is behind the imminent extinction of the vaquita, the world's smallest dolphin breed. Only found in Mexico's Gulf of California, the once numerous vaquita now number less than 60 in total. Vaquita die in gillnets set to catch totoaba (the world's largest drum fish). Totoaba are being hunted to extinction for its maw, which can sell for as much $10,000 per kilogram.

Swim bladders are also used in the food industry as a source of collagen. They can be made into a strong, water-resistant glue, or used to make isinglass for the clarification of beer. In earlier times they were used to make condoms.

Swim Bladder Disease

Swim bladder disease is a common ailment in aquarium fish. A fish with swim bladder disorder can float nose down tail up, or can float to the top or sink to the bottom of the aquarium.

Risk of Injury

Many anthropogenic activities, like pile driving or even Seismic wave, that could result from climate change or natural causes, can create high-intensity sound waves that cause a certain amount of damage to fishes that possess a gas bladder. Physostomes can release air in order to decrease the tension in the gas bladder that may cause internal injuries to other vital organs. While physoclisti can't expel air fast enough, making it more difficult to avoid any major injuries. Some of the commonly seen injuries included ruptured gas bladder and renal Haemorrhage. These mostly affect the overall health of the fish and didn't affect their mortality rate. Investigators used the High-Intensity-Controlled Impedance Fluid Filled (HICI-FT), a stainless-steel wave tube with a electromagnetic shaker. It simulates high-energy sound waves in aquatic far-field, plane-wave acoustic conditions.

Similar Structures in other Organisms

Siphonophores have a special swim bladder that allows the jellyfish-like colonies to float along the surface of the water while their tentacles trail below. This organ is unrelated to the one in fish.

Sensory System in Fish

Most fish possess highly developed sense organs. Nearly all daylight fish have color vision that is at least as good as a human's (see vision in fishes). Many fish also have chemoreceptors that are responsible for extraordinary senses of taste and smell. Although they have ears, many fish may not hear very well. Most fish have sensitive receptors that form the lateral line system, which detects gentle currents and vibrations, and senses the motion of nearby fish and prey. Sharks can sense frequencies in the range of 25 to 50 Hz through their lateral line.

Fish orient themselves using landmarks and may use mental maps based on multiple landmarks or symbols. Fish behavior in mazes reveals that they possess spatial memory and visual discrimination.

Vision

Fish have a refractive index gradient within the lens of their eyes which compensates for spherical aberration. Unlike humans, most fish adjust focus by moving the lens closer or further from the retina.

Vision is an important sensory system for most species of fish. Fish eyes are similar to those of terrestrial vertebrates like birds and mammals, but have a more spherical lens. Their retinas generally have both rod cells and cone cells (for scotopic and photopic vision), and most species have colour vision. Some fish can see ultraviolet and some can see polarized light. Amongst jawless fish, the lamprey has well-developed eyes, while the hagfish has only primitive eyespots. Fish vision shows adaptation to their visual environment, for example deep sea fishes have eyes suited to the dark environment.

Fish and other aquatic animals live in a different light environment than terrestrial species. Water absorbs light so that with increasing depth the amount of light available decreases quickly. The optic properties of water also lead to different wavelengths of light being absorbed to different degrees, for example light of long wavelengths (e.g. red, orange) is absorbed quite quickly compared to light of short wavelengths (blue, violet), though ultraviolet light (even shorter wavelength than blue) is absorbed quite quickly as well. Besides these universal qualities of water, different bodies of water may absorb light of different wavelengths because of salts and other chemicals in the water.

Hearing

Hearing is an important sensory system for most species of fish. Hearing threshold and the ability to localize sound sources are reduced underwater, in which the speed of sound is faster than in air. Underwater hearing is by bone conduction, and localization of sound appears to depend on differences in amplitude detected by bone conduction. Aquatic animals such as fish, however, have a more specialized hearing apparatus that is effective underwater.

Fish can sense sound through their lateral lines and their otoliths (ears). Some fishes, such as some species of carp and herring, hear through their swim bladders, which function rather like a hearing aid.

Hearing is well-developed in carp, which have the Weberian organ, three specialized vertebral processes that transfer vibrations in the swim bladder to the inner ear.

Although it is hard to test sharks' hearing, they may have a sharp sense of hearing and can possibly hear prey many miles away. A small opening on each side of their heads (not the spiracle) leads directly into the inner ear through a thin channel. The lateral line shows a similar arrangement, and is open to the environment via a series of openings called lateral line pores. This is a reminder of the common origin of these two vibration- and sound-detecting organs that are grouped together as the acoustico-lateralis system. In bony fish and tetrapods the external opening into the inner ear has been lost.

Current Detection

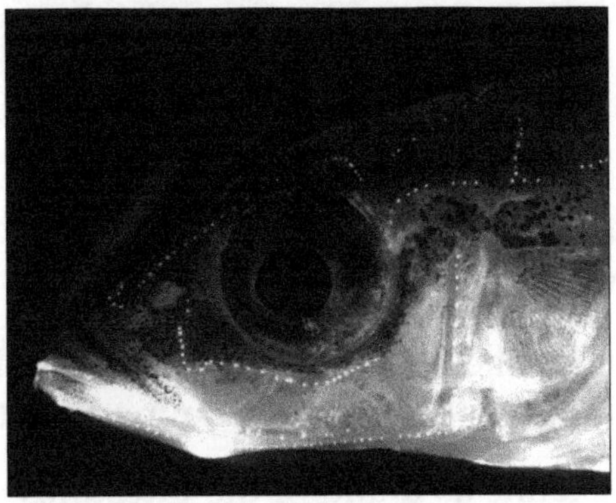

A *three-spined stickleback* with stained neuromasts.

Hair cells in fish are used to detect water movements around their bodies. These hair cells are embedded in a jelly-like protrusion called cupula. The hair cells therefore can not be seen and do not appear on the surface of skin.

The lateral line in fish and aquatic forms of amphibians is a detection system of water currents, consisting mostly of vortices. The lateral line is also sensitive to low-frequency vibrations. The mechanoreceptors are hair cells, the same mechanoreceptors for vestibular sense and hearing. It is used primarily for navigation, hunting, and schooling. The receptors of the electrical sense are modified hair cells of the lateral line system.

Fish and some aquatic amphibians detect hydrodynamic stimuli via a lateral line. This system consists of an array of sensors called neuromasts along the length of the fish's body. Neuromasts can be free-standing (superficial neuromasts) or within fluid-filled canals (canal neuromasts). The sensory cells within neuromasts are polarized hair cells contained within a gelatinous cupula. The cupula, and the stereocilia within, are moved by a certain amount depending on the movement of the surrounding water. Afferent nerve fibers are excited or inhibited depending on whether the hair cells they arise from are deflected in the preferred or opposite direction. Lateral line receptors form somatotopic maps within the brain informing the fish of amplitude and direction of flow at different points along the body. These maps are located in the medial octavolateral nucleus (MON) of the medulla and in higher areas such as the torus semicircularis.

Pressure Detection

Pressure detection uses the organ of Weber, a system consisting of three appendages of vertebrae transferring changes in shape of the gas bladder to the middle ear. It can be used to regulate the buoyancy of the fish. Fish like the weather fish and other loaches are also known to respond to low pressure areas but they lack a swim bladder.

Chemoreception

The shape of the hammerhead shark's head may enhance olfaction by spacing the nostrils further apart.

Sharks have keen olfactory senses, located in the short duct (which is not fused, unlike bony fish) between the anterior and posterior nasal openings, with some species able to detect as little as one part per million of blood in seawater.

Sharks have the ability to determine the direction of a given scent based on the timing of scent detection in each nostril. This is similar to the method mammals use to determine direction of sound.

They are more attracted to the chemicals found in the intestines of many species, and as a result often linger near or in sewage outfalls. Some species, such as nurse sharks, have external barbels that greatly increase their ability to sense prey.

The MHC genes are a group of genes present in many animals and important for the immune system; in general, offspring from parents with differing MHC genes have a stronger immune system.

Fish are able to smell some aspect of the MHC genes of potential sex partners and prefer partners with MHC genes different from their own.

Salmon have a strong sense of smell. Speculation about whether odours provide homing cues, go back to the 19th century. In 1951, Hasler hypothesised that, once in vicinity of the estuary or entrance to its birth river, salmon may use chemical cues which they can smell, and which are unique to their natal stream, as a mechanism to home onto the entrance of the stream. In 1978, Hasler and his students convincingly showed that the way salmon locate their home rivers with such precision was indeed because they could recognise its characteristic smell. They further demonstrated that the smell of their river becomes imprinted in salmon when they transform into smolts, just before they migrate out to sea. Homecoming salmon can also recognise characteristic smells in tributary streams as they move up the main river. They may also be sensitive to characteristic pheromones given off by juvenile conspecifics. There is evidence that they can "discriminate between two populations of their own species".

Electroreception and Magnetoreception

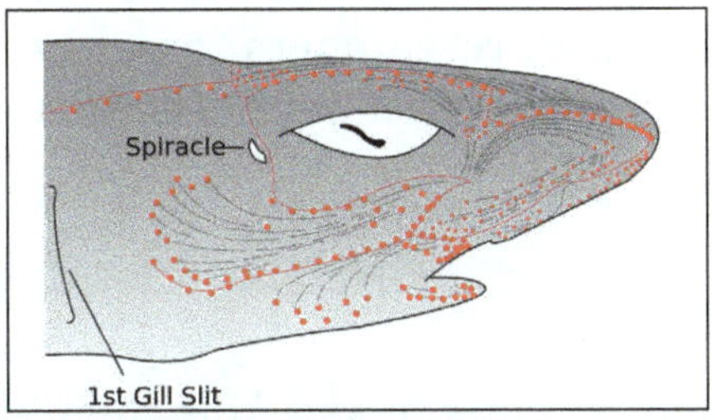

Electromagnetic field receptors (ampullae of Lorenzini) and motion detecting canals in the head of a shark.

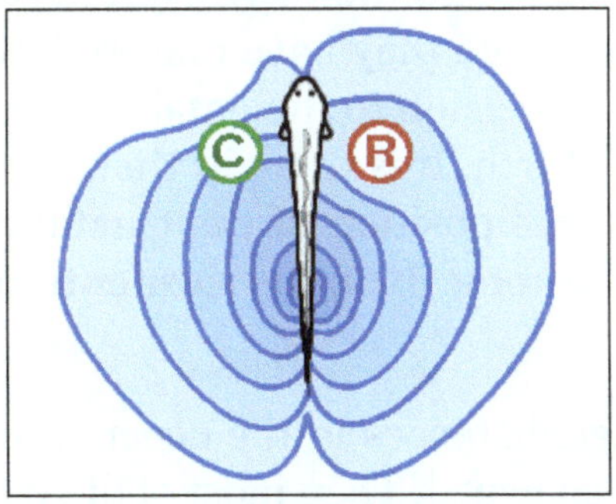

Active electrolocation. Conductive objects concentrate the field and resistive objects spread the field.

Electroreception, or electroception, is the ability to detect electric fields or currents. Some fish, such as catfish and sharks, have organs that detect weak electric potentials on the order of millivolts. Other fish, like the South American electric fishes Gymnotiformes, can produce weak electric currents, which they use in navigation and social communication. In sharks, the ampullae of Lorenzini are electroreceptor organs. They number in the hundreds to thousands. Sharks use the ampullae of Lorenzini to detect the electromagnetic fields that all living things produce. This helps sharks (particularly the hammerhead shark) find prey. The shark has the greatest electrical sensitivity of any animal. Sharks find prey hidden in sand by detecting the electric fields they produce. Ocean currents moving in the magnetic field of the Earth also generate electric fields that sharks can use for orientation and possibly navigation.

Electric field proximity sensing is used by the electric catfish to navigate through muddy waters. These fish make use of spectral changes and amplitude modulation to determine factors such shape, size, distance, velocity, and conductivity. The abilities of the electric fish to communicate and identify sex, age, and hierarchy within the species are also made possible through electric fields. EF gradients as low as 5nV/cm can be found in some saltwater weakly electric fish.

The paddlefish (*Polyodon spathula*) hunts plankton using thousands of tiny passive electroreceptors located on its extended snout, or rostrum. The paddlefish is able to detect electric fields that oscillate at 0.5–20 Hz, and large groups of plankton generate this type of signal.

Electric fishes use an active sensory system to probe the environment and create active electrodynamic imaging.

In 1973, it was shown that Atlantic salmon have conditioned cardiac responses to electric fields with strengths similar to those found in oceans. "This sensitivity might allow a migrating fish to align itself upstream or downstream in an ocean current in the absence of fixed references."

Magnetoception, or magnetoreception, is the ability to detect the direction one is facing based on the Earth's magnetic field. In 1988, researchers found iron, in the form of single domain magnetite, resides in the skulls of sockeye salmon. The quantities present are sufficient for magnetoception.

Fish Navigation

Salmon regularly migrate thousands of miles to and from their breeding grounds.

Salmon spend their early life in rivers, and then swim out to sea where they live their adult lives and gain most of their body mass. When they have matured, they return to the rivers to spawn. Usually they return with uncanny precision to the natal river where they were born, and even to the very spawning ground of their birth. It is thought that, when they are in the ocean, they use magnetoception to locate the general position of their natal river, and once close to the river, that they use their sense of smell to home in on the river entrance and even their natal spawning ground.

After several years wandering huge distances in the ocean, most surviving salmon return to the same natal rivers where they were spawned. Then most of them swim up the rivers until they reach the very spawning ground that was their original birthplace.

There are various theories about how this happens. One theory is that there are geomagnetic and chemical cues which the salmon use to guide them back to their birthplace. The fish may be sensitive to the Earth's magnetic field, which could allow the fish to orient itself in the ocean, so it can navigate back to the estuary of its natal stream.

Pain

Experiments done by William Tavolga provide evidence that fish have pain and fear responses. For instance, in Tavolga's experiments, toadfish grunted when electrically shocked and over time they came to grunt at the mere sight of an electrode.

Hooked sailfish

In 2003, Scottish scientists at the University of Edinburgh and the Roslin Institute concluded that rainbow trout exhibit behaviors often associated with pain in other animals. Bee venom and acetic acid injected into the lips resulted in fish rocking their bodies and rubbing their lips along the sides and floors of their tanks, which the researchers concluded were attempts to relieve pain, similar to what mammals would do. Neurons fired in a pattern resembling human neuronal patterns.

Professor James D. Rose of the University of Wyoming claimed the study was flawed since it did not provide proof that fish possess "conscious awareness, particularly a kind of awareness that is meaningfully like ours". Rose argues that since fish brains are so different from human brains, fish are probably not conscious in the manner humans are, so that reactions similar to human reactions to pain instead have other causes. Rose had published a study a year earlier arguing that fish cannot feel pain because their brains lack a neocortex. However, animal behaviorist Temple Grandin argues that fish could still have consciousness without a neocortex because "different species can use different brain structures and systems to handle the same functions."

Animal welfare advocates raise concerns about the possible suffering of fish caused by angling. Some countries, such as Germany have banned specific types of fishing, and the British RSPCA now prosecutes individuals who are cruel to fish.

Pain in Fish

Whether fish, such as this hooked salmon, can be said to feel pain is controversial.

Pain in fish is a contentious issue. Pain is a complex mental state, with a distinct perceptual quality but also associated with suffering, which is an emotional state. Because of this complexity, the presence of pain in an animal, or another human for that matter, cannot be determined unambiguously using observational methods, but the conclusion that animals experience pain is often inferred on the basis of likely presence of phenomenal consciousness which is deduced from comparative brain physiology as well as physical and behavioural reactions.

Fish fulfill several criteria proposed as indicating that non-human animals may experience pain. These fulfilled criteria include a suitable nervous system and sensory receptors, opioid receptors and reduced responses to noxious stimuli when given analgesics and local anaesthetics, physiological changes to noxious stimuli, displaying protective motor reactions, exhibiting avoidance learning and making trade-offs between noxious stimulus avoidance and other motivational requirements.

If fish feel pain, there are ethical and animal welfare implications including the consequences of exposure to pollutants, and practices involving commercial and recreational fishing, aquaculture, in ornamental fish and genetically modified fish and for fish used in scientific research.

Background

The possibility that fish and other non-human animals may experience pain has a long history. Initially, this was based around theoretical and philosophical argument, but more recently has turned to scientific investigation.

Philosophy

René Descartes

The idea that non-human animals might not feel pain goes back to the 17th-century French philosopher, René Descartes, who argued that animals do not experience pain and suffering because they lack consciousness. In 1789, the British philosopher and social reformist, Jeremy Bentham, addressed in his book *An Introduction to the Principles of Morals and Legislation* the issue of our treatment of animals with the following often quoted words: "The question is not, Can they reason? nor, can they talk? but, Can they suffer?"

Peter Singer, a bioethicist and author of *Animal Liberation* published in 1975, suggested that consciousness is not necessarily the key issue: just because animals have smaller brains, or are 'less conscious' than humans, does not mean that they are not capable of feeling pain. He goes on further to argue that we do not assume newborn infants, people suffering from neurodegenerative brain diseases or people with learning disabilities experience less pain than we would.

Bernard Rollin, the principal author of two U.S. federal laws regulating pain relief for animals, writes that researchers remained unsure into the 1980s as to whether animals experience pain, and veterinarians trained in the U.S. before 1989 were taught to simply ignore animal pain. In his interactions with scientists and other veterinarians, Rollin was regularly asked to "prove" that animals are conscious, and to provide "scientifically acceptable" grounds for claiming that they feel pain.

Continuing into the 1990s, discussions were further developed on the roles that philosophy and science had in understanding animal cognition and mentality. In subsequent years, it was argued there was strong support for the suggestion that some animals (most likely amniotes) have at least simple conscious thoughts and feelings and that the view animals feel pain differently to humans is now a minority view.

Scientific Investigation

Cambridge Declaration on Consciousness (2012)

The absence of a neocortex does not appear to preclude an organism from experiencing affective states. Convergent evidence indicates that non-human animals have the neuroanatomical, neurochemical, and neurophysiological substrates of conscious states along with the capacity to exhibit intentional behaviors. Consequently, the weight of evidence indicates that humans are not unique in possessing the neurological substrates that generate consciousness. Non-human animals, including all mammals and birds, and many other creatures, including octopuses, also possess these neurological substrates.

In the 20th and 21st centuries, there were many scientific investigations of pain in non-human animals.

Mammals

At the turn of the century, studies were published showing that arthritic rats self-select analgesic opiates. In 2014, the veterinary *Journal of Small Animal Practice* published an article on the recognition of pain which started – "The ability to experience pain is universally shared by all mammals..." and in 2015, it was reported in the science journal *Pain*, that several mammalian species (rat, mouse, rabbit, cat and horse) adopt a facial expression in response to a noxious stimulus that is consistent with the expression of humans in pain.

Birds

At the same time as the investigations using arthritic rats, studies were published showing that birds with gait abnormalities self-select for a diet that contains carprofen, a human analgesic. In 2005, it was written "Avian pain is likely analogous to pain experienced by most mammals" and in 2014, "...it is accepted that birds perceive and respond to noxious stimuli and that birds feel pain".

Reptiles and Amphibians

Veterinary articles have been published stating both reptiles and amphibians experience pain in a way analogous to humans, and that analgesics are effective in these two classes of vertebrates.

Argument by Analogy

In 2012 the American philosopher Gary Varner reviewed the research literature on pain in animals. His findings are summarised in the following table.

Argument by analogy					
Property	Fish	Amphibians	Reptiles	Birds	Mammals
Has nociceptors	✓	✓	✓	✓	✓
Has brain	✓	✓	✓	✓	✓
Nociceptors and brain linked	✓	✓	✓	✓	✓
Has endogenous opioids	✓	✓	✓	✓	✓
Analgesics affect responses	✓			✓	✓
Response to damaging stimuli similar to humans	✓	✓	✓	✓	✓

Arguing by analogy, Varner claims that any animal which exhibits the properties listed in the table could be said to experience pain. On that basis, he concludes that all vertebrates, including fish, probably experience pain, but invertebrates apart from cephalopods probably do not experience pain.

The Experience of Pain

Although there are numerous definitions of pain, almost all involve two key components.

First, nociception is required. This is the ability to detect noxious stimuli which evoke a reflex response that rapidly moves the entire animal, or the affected part of its body, away from the source of the stimulus. The concept of nociception does not imply any adverse, subjective "feeling" – it is a reflex action. An example in humans would be the rapid withdrawal of a finger that has touched something hot – the withdrawal occurs before any sensation of pain is actually experienced.

The second component is the experience of "pain" itself, or suffering – the internal, emotional interpretation of the nociceptive experience. Again in humans, this is when the withdrawn finger begins to hurt, moments after the withdrawal. Pain is therefore a private, emotional experience. Pain cannot be directly measured in other animals, including other humans; responses to putatively painful stimuli can be measured, but not the experience itself. To address this problem when assessing the capacity of other species to experience pain, argument-by-analogy is used. This is based on the principle that if an animal responds to a stimulus in a similar way to ourselves, it is likely to have had an analogous experience.

Nociception

Nociception usually involves the transmission of a signal along a chain of nerve fibers from the site of a noxious stimulus at the periphery to the spinal cord and brain. This process evokes a reflex arc response generated at the spinal cord and not involving the brain, such as flinching or withdrawal of a limb. Nociception is found, in one form or another, across all major animal taxa. Nociception can be observed using modern imaging techniques; and a physiological and behavioral response to nociception can often be detected. However, nociceptive responses can be so subtle in prey animals that trained (human) observers can not perceive them, whereas natural predators can and subsequently target injured individuals.

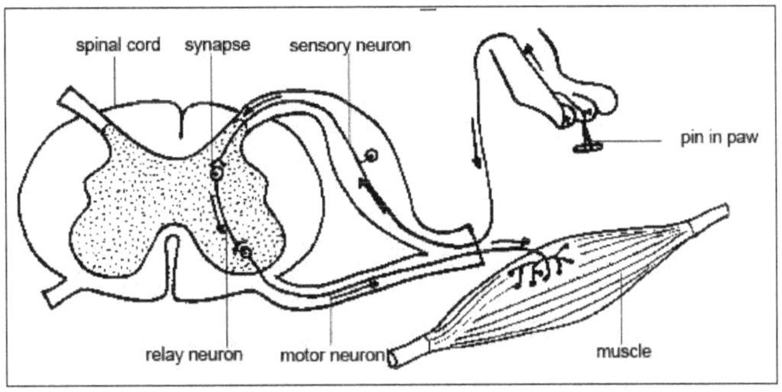

Nociception: The reflex arc of a dog with a pin in her paw. Note there is no communication to the brain, but the paw is withdrawn by nervous impulses generated by the spinal cord. There is no conscious interpretation of the stimulus by the dog.

Emotional Pain

Sometimes a distinction is made between "physical pain" and "emotional" or "psychological pain". Emotional pain is the pain experienced in the absence of physical trauma, e.g. the pain experienced by humans after the loss of a loved one, or the break-up of a relationship. It has been argued that only primates and humans can feel "emotional pain", because they are the only animals that have a neocortex – a part of the brain's cortex considered to be the "thinking area". However, research has provided evidence that monkeys, dogs, cats and birds can show signs of emotional pain and display behaviours associated with depression during painful experience, i.e. lack of motivation, lethargy, anorexia, unresponsiveness to other animals.

Physical Pain

The nerve impulses of the nociception response may be conducted to the brain thereby registering the location, intensity, quality and unpleasantness of the stimulus. This subjective component of pain involves conscious awareness of both the sensation and the unpleasantness (the aversive, negative affect). The brain processes underlying conscious awareness of the unpleasantness (suffering), are not well understood.

There have been several published lists of criteria for establishing whether non-human animals experience pain, e.g. Some criteria that may indicate the potential of another species, including fishes, to feel pain include:

1. Has a suitable nervous system and sensory receptors.
2. Has opioid receptors and shows reduced responses to noxious stimuli when given analgesics and local anaesthetics.

3. Physiological changes to noxious stimuli.
4. Displays protective motor reactions that might include reduced use of an affected area such as limping, rubbing, holding or autotomy.
5. Shows avoidance learning.
6. Shows trade-offs between noxious stimulus avoidance and other motivational requirements.
7. High cognitive ability and sentience.

Adaptive Value

The adaptive value of nociception is obvious; an organism detecting a noxious stimulus immediately withdraws the limb, appendage or entire body from the noxious stimulus and thereby avoids further (potential) injury. However, a characteristic of pain (in mammals at least) is that pain can result in hyperalgesia (a heightened sensitivity to noxious stimuli) and allodynia (a heightened sensitivity to non-noxious stimuli). When this heightened sensitisation occurs, the adaptive value is less clear. First, the pain arising from the heightened sensitisation can be disproportionate to the actual tissue damage caused. Second, the heightened sensitisation may also become chronic, persisting well beyond the tissues healing. This can mean that rather than the actual tissue damage causing pain, it is the pain due to the heightened sensitisation that becomes the concern. This means the sensitisation process is sometimes termed maladaptive. It is often suggested hyperalgesia and allodynia assist organisms to protect themselves during healing, but experimental evidence to support this has been lacking.

In 2014, the adaptive value of sensitisation due to injury was tested using the predatory interactions between longfin inshore squid (*Doryteuthis pealeii*) and black sea bass (*Centropristis striata*) which are natural predators of this squid. If injured squid are targeted by a bass, they began their defensive behaviours sooner (indicated by greater alert distances and longer flight initiation distances) than uninjured squid. If anaesthetic (1% ethanol and $MgCl_2$) is administered prior to the injury, this prevents the sensitisation and blocks the behavioural effect. The authors claim this study is the first experimental evidence to support the argument that nociceptive sensitisation is actually an adaptive response to injuries.

The question has been asked, “If fish cannot feel pain, why do stingrays have purely defensive tail spines that deliver venom? Stingrays’ ancestral predators are fish. And why do many fishes possess defensive fin spines, some also with venom that produces pain in humans?”

Peripheral Nervous System

Receptors

Primitive fish such as lampreys (*Petromyzon marinus*) have free nerve endings in the skin that respond to heat and mechanical pressure. However, behavioural reactions associated with nociception have not been recorded and it is also difficult to determine whether the mechanoreceptors in lamprey are truly nociceptive-specific or simply pressure-specific.

Rainbow trout have nociceptors on the face, snout and other areas of the body.

Rainbow trout (*Oncorhynchus mykiss*) have polymodal nociceptors on the face and snout which respond to mechanical pressure, temperatures in the noxious range (> 40°C) and 1% acetic acid (a chemical irritant). Further studies found nociceptors to be more widely distributed over the bodies of rainbow trout, and also those of cod and carp. The most sensitive areas of the body are around the eyes, nostrils, fleshy parts of the tail and the pectoral and dorsal fins.

Rainbow trout also have corneal nociceptors. Out of 27 receptors investigated in one study, seven were polymodal nociceptors and six were mechanothermal nociceptors. Mechanical and thermal thresholds were lower than those of cutaneous receptors, indicating greater sensitivity in the cornea.

Bony fish possess nociceptors that are similar in function to those in mammals.

Nerve Fibres

There are two types of nerve fibre relevant to pain in fish. Group C nerve fibres are a type of sensory nerve fibre which lack a myelin sheath and have a small diameter, meaning they have a low nerve conduction velocity. The suffering that humans associate with burns, toothaches, or crushing injury are caused by C fibre activity. A typical human cutaneous nerve contains 83% Group C nerve fibres. A-delta fibres are another type of sensory nerve fibre, however, these are myelinated and therefore transmit impulses faster than non-myelinated C fibres. A-delta fibres carry cold, pressure and some pain signals, and are associated with acute pain that results in "pulling away" from noxious stimuli.

Bony fish possess both Group C and A-delta fibres representing 38.7% (combined) of the fibres in the tail nerves of common carp and 36% of the trigeminal nerve of rainbow trout. However, only 5% and 4% of these are C fibres in the carp and rainbow trout, respectively.

Some species of cartilagenous fish possess A-delta fibres, however, C fibres are either absent or found in very low numbers. The Agnatha (hagfishes and lamprey) primarily have Group C fibres.

Central Nervous System

The central nervous system (CNS) of fish contains a spinal cord, medulla oblongata, and the brain, divided into telencephalon, diencephalon, mesencephalon and cerebellum.

In fish, similar to other vertebrates, nociception travels from the peripheral nerves along the spinal nerves and is relayed through the spinal cord to the thalamus. The thalamus is connected to the telencephalon by multiple connections through the grey matter pallium, which has been demonstrated to receive nerve relays for noxious and mechanical stimuli.

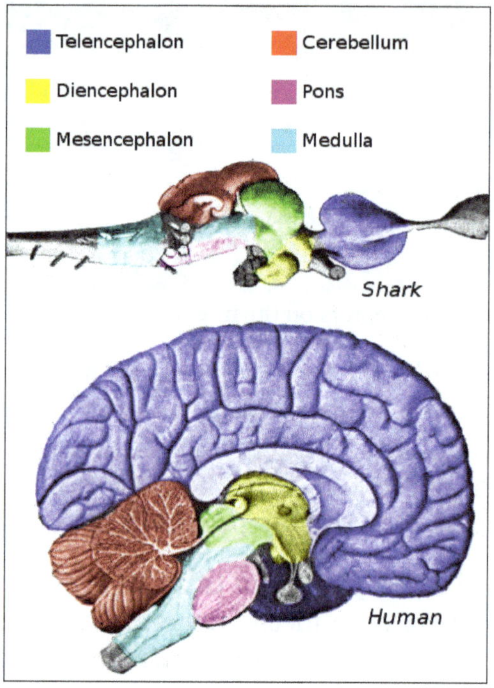

The brain regions of sharks and humans.

The major tracts that convey pain information from the periphery to the brain are the spinothalamic tract (body) and the trigeminal tract (head). Both have been studied in agnathans, teleost, and elasmobranch fish (trigeminal in the common carp, spinothalamic tract in the sea robin, *Prionotus carolinus*).

Brain

If sensory responses in fish are limited to the spinal cord and hindbrain, they might be considered as simply reflexive. However, recordings from the spinal cord, cerebellum, tectum and telencephalon in both trout and goldfish (*Carassius auratus*) show these all respond to noxious stimuli. This indicates a nociceptive pathway from the periphery to the higher CNS of fish.

Microarray analysis of gene expression shows the brain is active at the molecular level in the forebrain, midbrain and hindbrain of common carp and rainbow trout. Several genes involved in mammalian nociception, such as brain-derived neurotrophic factor (BDNF) and the cannabinoid CB1 receptor are regulated in the fish brain after a nociceptive event.

Somatosensory evoked potentials (SEPs) are weak electric responses in the CNS following stimulation of peripheral sensory nerves. These further indicate there is a pathway from the peripheral nociceptors to higher brain regions. In goldfish, rainbow trout, Atlantic salmon (*Salmo salar*) and Atlantic cod (*Gadus morhua*), it has been demonstrated that putatively non-noxious and noxious stimulation elicit SEPs in different brain regions, including the telencephalon which may mediate the co-ordination of pain information. Moreover, multiple functional magnetic resonance imaging (fMRI) studies with several species of fishes have shown that when suffering from putative pain, there is profound activity in the forebrain which is highly reminiscent of that observed in humans and would be taken as evidence of the experience of pain in mammals.

Therefore, "higher" brain areas are activated at the molecular, physiological, and functional levels in fish experiencing a potentially painful event. Sneddon stated "This gives much weight to the proposal that fish experience some form of pain rather than a nociceptive event".

Opioid System and Effects of Analgesics

Analgesics and anaesthetics are commonly used for surgery on fish.

Teleost fish have a functional opioid system which includes the presence of opioid receptors similar to those of mammals. All four of the main opioid receptor types (delta, kappa, mu, and NOP) are conserved in vertebrates, even in primitive jawless fishes (agnathastoma).

The same analgesics and anaesthetics used in humans and other mammals, are often used for fish in veterinary medicine. These chemicals act on the nociceptive pathways, blocking signals to the brain where emotional responses to the signals are further processed by certain parts of the brain found in amniotes ("higher vertebrates").

Effects of Morphine

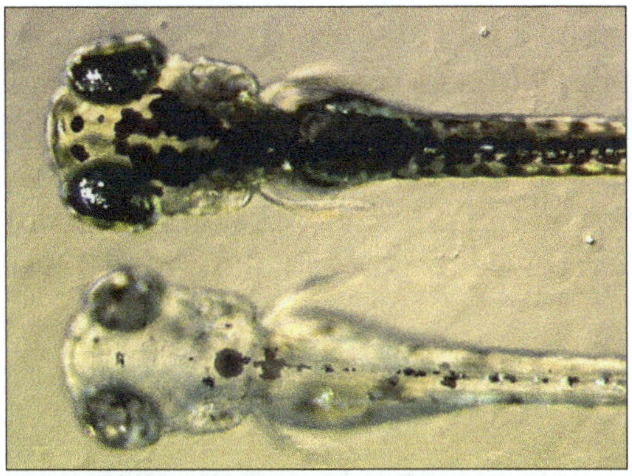

Five-day-old zebrafish larvae show behavioural responses indicative of pain in response to diluted acetic acid.

Pre-treatment with morphine (an analgesic in humans and other mammals) has a dose-dependent anti-nociceptive effect and mitigates the behavioural and ventilation rate responses of rainbow trout to noxious stimuli.

When acetic acid is injected into the lips of rainbow trout, they exhibit anomalous behaviours such as side-to-side rocking and rubbing their lips along the sides and floors of the tanks, and their ventilation rate increases. Injections of morphine reduce both the anomalous, noxious-stimulus related behaviours and the increase in ventilation rate. When the same noxious stimulus is applied to zebrafish (*Danio rerio*), they respond by decreasing their activity. As with the rainbow trout, morphine injected prior to the acid injection attenuates the decrease in activity in a dose-dependent manner.

Injection of acetic acid into the lips of rainbow trout causes a reduction in their natural neophobia (fear of novelty); this is reversed by the administration of morphine.

In goldfish injected with morphine or saline and then exposed to unpleasant temperatures, fish injected with saline acted with defensive behaviours indicating anxiety, wariness and fear, whereas those given morphine did not.

Effects of other Analgesics

The neurotransmitter, Substance P and the analgesic opioid enkephalins and β-endorphin, which act as endogenous analgesics in mammals, are present in fish.

Different analgesics have different effects on fish. In a study on the efficacy of three types of analgesic, buprenorphine (an opioid), carprofen (a non-steroidal anti-inflammatory drug) and lidocaine (a local anaesthetic), ventilation rate and time to resume feeding were used as pain indicators. Buprenorphine had limited impact on the fish's response, carprofen ameliorated the effects of noxious stimulation on time to resume feeding, however, lidocaine reduced all the behavioural indicators.

Tramadol also increases the nociceptive threshold in fish, providing further evidence of an anti-nociceptive opioid system in fish.

Effects of Naloxone

Naloxone is an μ-opioid receptor antagonist which, in mammals, negates the analgesic effects of opioids. Both adult and 5-day old zebrafish larvae show behavioural responses indicative of pain in response to injected or diluted acetic acid. The anti-nociceptive properties of morphine or buprenorphine are reversed if adults, or larvae, are co-treated with naloxone. Both naloxone and prolyl-leucyl-glycinamide (another opiate antagonist in mammals) reduced the analgesic effects of morphine to electric shocks received by goldfish, indicating they can act as an opiate antagonist in fish.

Physiological Changes

The physiological changes of fish in response to noxious stimuli include elevations of ventilation rate and cortisol levels.

Protective Responses

Studies show that fish exhibit protective behavioural responses to putatively painful stimuli.

Noxiously stimulated common carp show anomalous rocking behaviour and rub their lips against the tank walls.

Noxiously stimulated zebrafish reduce their frequency of swimming and increase their ventilation rate.

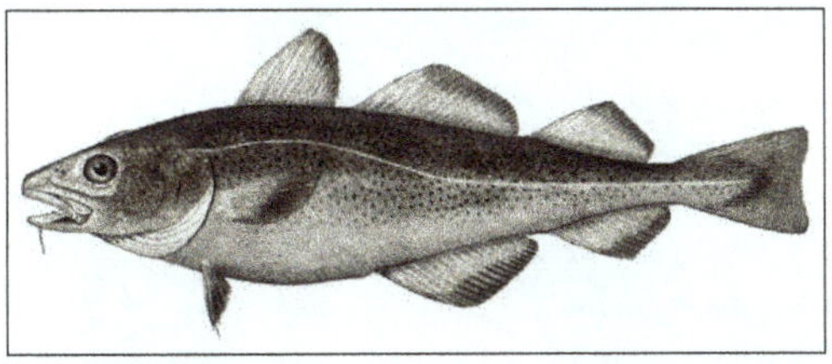

Noxiously stimulated Atlantic cod display increased hovering close to the bottom of the tank and reduced use of shelter.

When acetic acid or bee venom is injected into the lips of rainbow trout, they exhibit an anomalous side-to-side rocking behaviour on their pectoral fins, rub their lips along the sides and floors of the tanks and increase their ventilation rate. When acetic acid is injected into the lips of zebrafish, they respond by decreasing their activity. The magnitude of this behavioural response depends on the concentration of the acetic acid.

The behavioural responses to a noxious stimulus differ between species of fish. Noxiously stimulated common carp (*Cyprinus carpio*) show anomalous rocking behaviour and rub their lips against the tank walls, but do not change other behaviours or their ventilation rate. In contrast, zebrafish (*Danio rerio*) reduce their frequency of swimming and increase their ventilation rate but do not display anomalous behaviour. Rainbow trout, like the zebrafish, reduce their frequency of swimming and increase their ventilation rate. Nile tilapia (*Oreochromis niloticus*), in response to a tail fin clip, increase their swimming activity and spend more time in the light area of their tank.

Since this initial work, Sneddon and her co-workers have shown that rainbow trout, common carp and zebrafish experiencing a noxious stimulation exhibit rapid changes in physiology and behavior that persist for up to 6 hours and thus are not simple reflexes.

Five-day old zebrafish larvae show a concentration dependent increase in locomotor activity in response to different concentrations of diluted acetic acid. This increase in locomotor activity is accompanied by an increase in cox-2 mRNA, demonstrating that nociceptive pathways are also activated.

Fish show different responses to different noxious stimuli, even when these are apparently similar. This indicates the response is flexible and not simply a nociceptive reflex. Atlantic cod injected in the lip with acetic acid, capsaicin, or piercing the lip with a commercial fishing hook, showed different responses to these three types of noxious stimulation. Those cod treated with acetic acid and capsaicin displayed increased hovering close to the bottom of the tank and reduced use of shelter. However, hooked cod only showed brief episodes of head shaking.

Avoidance Learning

Early experiments provided evidence that fish learn to respond to putatively noxious stimuli. For instance, toadfish (*Batrachoididae*) grunt when they are electrically shocked, but after repeated shocks, they grunt simply at the sight of the electrode. More recent studies show that both goldfish and trout learn to avoid locations in which they receive electric shocks. Furthermore, this avoidance learning is flexible and is related to the intensity of the stimulus.

Trade-offs in Motivation

Goldfish make trade-offs between their motivation to feed or avoid an acute noxious stimulus.

A painful experience may change the motivation for normal behavioural responses.

In a 2007 study, goldfish were trained to feed at a location of the aquarium where subsequently they would receive an electric shock. The number of feeding attempts and time spent in the feeding/shock zone decreased with increased shock intensity and with increased food deprivation the number and the duration of feeding attempts increased as did escape responses as this zone was entered. The researchers suggested that goldfish make a trade-off in their motivation to feed with their motivation to avoid an acute noxious stimulus.

Rainbow trout naturally avoid novelty (i.e. they are neophobic). Braithwaite describes a study in which a brightly coloured Lego brick is placed in the tank of rainbow trout. Trout injected in the lip with a small amount of saline strongly avoided the Lego brick, however, trout injected with acetic acid spent considerably more time near the Lego block. When the study was repeated but with the fish also being given morphine, the avoidance response returned in those fish injected with acetic acid and could not be distinguished from the responses of saline injected fish.

To explore the possibility of a trade-off between responding to a noxious stimulus and predation, researchers presented rainbow trout with a competing stimulus, a predator cue. Noxiously stimulated fish cease showing anti-predator responses, indicating that pain becomes their primary motivation. The same study investigated the potential trade-off between responding to a noxious stimulus and social status. The responses of the noxiously treated trout varied depending on the familiarity of the fish they were placed with. The researchers suggested the findings of the motivational changes and trade-offs provide evidence for central processing of pain rather than merely showing a nociceptive reflex.

Paying a Cost for Analgesia

Zebrafish given access to a barren, brightly lit chamber or an enriched chamber prefer the enriched area. When these fish are injected with acetic acid or saline as a control they still choose the same enriched chamber. However, if an analgesic is dissolved in the barren, less-preferred chamber, zebrafish injected with noxious acid lose their preference and spend over half their time in the previously less-favourable, barren chamber. This suggests a trade-off in motivation and furthermore, they are willing to pay a cost to enter a less preferred environment to access pain relief.

Cognitive Ability and Sentience

The learning abilities of fish demonstrated in a range of studies indicate sophisticated cognitive processes that are more complex than simple associative learning. Examples include the ability to recognise social companions, avoidance (for some months or years) of places where they encountered a predator or were caught on a hook and forming mental maps.

It has been argued that although a high cognitive capacity may indicate a greater likelihood of experiencing pain, it also gives these animals a greater ability to deal with this, leaving animals with a lower cognitive ability a greater problem in coping with pain.

Societal Implications

There are concerns that angling causes pain in fish.

Given the broad scientific consensus that fish can probably feel pain, it has been suggested that precautionary principles should be applied to commercial fishing, which would likely have multiple consequences.

Both scientists and animal protection advocates have raised concerns about the possible suffering (pain and fear) of fish caused by angling.

Other societal implications of fish experiencing pain include acute and chronic exposure to pollutants, commercial and sporting fisheries (e.g. injury during trawling, tagging/fin clipping during stock assessment, tissue damage, physical exhaustion and severe oxygen deficit during capture, pain and stress during slaughter, use of live bait), aquaculture (e.g. tagging/fin clipping, high stocking densities resulting in increased aggression, food deprivation for disease treatment or before harvest, removal from water for routine husbandry, pain during slaughter), ornamental fish (e.g. capture by sub-lethal poisoning, permanent adverse physical states due to selective breeding), scientific research (e.g. genetic-modification) may have detrimental effects on welfare, deliberately-imposed adverse physical, physiological and behavioural states, electrofishing, tagging, fin clipping or otherwise marking fish, handling procedures which may cause injury.

Legislation

In the UK, the legislation protecting animals during scientific research, the "Animals (Scientific Procedures) Act 1986", protects fish from the moment they become capable of independent feeding. The legislation protecting animals in most other circumstances in the UK is "The Animal Welfare Act, 2006" which states that in the Act, " "animal" means a vertebrate other than man", clearly including fish.

In the US, the legislation protecting animals during scientific research is "The Animal Welfare Act". This excludes protection of "cold-blooded" animals, including fish.

The 1974 Norwegian Animal Rights Law states it relates to mammals, birds, frogs, salamander, reptiles, fish, and crustaceans.

Controversy

Receptors and Nerve Fibres

It has been argued that fish can not feel pain because they do not have a sufficient density of appropriate nerve fibres. A typical human cutaneous nerve contains 83% Group C nerve fibres, however, the same nerves in humans with congenital insensitivity to pain have only 24-28% C type fibres. Based on this, James Rose, from the University of Wyoming, has argued that the absence of C type fibres in cartilagenous sharks and rays indicates that signalling leading to pain perception is likely to be impossible, and the low numbers for bony fish (e.g. 5% for carp and trout) indicate this is also highly unlikely for these fish. Rose concludes there is little evidence that sharks and rays possess the nociceptors required to initiate pain detection in the brain, and that, while bony fish are able to unconsciously learn to avoid injurious stimuli, they are little more likely to experience conscious pain than sharks.

A-delta type fibres, believed to trigger avoidance reactions, are common in bony fish, although they have not been found in sharks or rays.

Rose concludes that fishes have survived well in an evolutionary sense without the full range of nociception typical of humans or other mammals.

Brain

In 2002, Rose published reviews arguing that fish cannot feel pain because they lack a neocortex in the brain. (This would also rule out pain perception in most mammals, all birds and reptiles.) However, in 2003, a research team led by Lynne Sneddon concluded that the brains of rainbow trout fire neurons in the same way human brains do when experiencing pain. Rose criticized the study, claiming it was flawed, mainly because it did not provide proof that fish possess "conscious awareness, particularly a kind of awareness that is meaningfully like ours".

Rose, and more recently Brian Key from The University of Queensland, argue that because the fish brain is very different to humans, fish are probably not conscious in the manner humans are, and while fish may react in a way similar to the way humans react to pain, the reactions in the case of fish have other causes. Studies indicating that fish can feel pain were confusing nociception with feeling pain, says Rose. "Pain is predicated on awareness. The key issue is the distinction between nociception and pain. A person who is anaesthetised in an operating theatre will still respond physically to an external stimulus, but he or she will not feel pain." According to Rose and Key, the literature relating to the question of consciousness in fish is prone to anthropomorphisms and care is needed to avoid erroneously attributing human-like capabilities to fish. Sneddon suggests it is entirely possible that a species with a different evolutionary path could evolve different neural systems to perform the same functions (i.e. convergent evolution), as studies on the brains of birds have shown. Key agrees that phenomenal consciousness is likely to occur in mammals and birds, but not in fish. Animal behaviouralist Temple Grandin argues that fish could still have consciousness without a neocortex because "different species can use different brain structures and systems to handle the same functions." Sneddon proposes that to suggest a function suddenly arises without a primitive form defies the laws of evolution.

Other researchers also believe that animal consciousness does not require a neocortex, but can arise from homologous subcortical brain networks. It has been suggested that brainstem circuits can generate pain. This includes research with anencephalic children who, despite missing large portions of their cortex, express emotions. There is also evidence from activation studies showing brainstem mediated feelings in normal humans and foetal withdrawal responses to noxious stimulation but prior to development of the cortex.

Protective Responses

Initial work by Sneddon and her co-workers characterised behavioural responses in rainbow trout, common carp and zebrafish. However, when these experiments were repeated by Newby and Stevens without anaesthetic, rocking and rubbing behaviour was not observed, suggesting that some of the alleged pain responses observed by Sneddon and co-workers were likely to be due to recovery of the fish from anaesthesia.

Several researchers argue about the definition of pain used in behavioural studies, as the observations recorded were contradictory, non-validated and non-repeatable by other researchers. In 2012, Rose argued that fishes resume "normal feeding and activity immediately or soon after surgery".

Nordgreen said that the behavioural differences they found in response to uncomfortable temperatures showed that fish feel both reflexive and cognitive pain. "The experiment shows that fish do not only respond to painful stimuli with reflexes, but change their behavior also after the event," Nordgreen said. "Together with what we know from experiments carried out by other groups, this indicates that the fish consciously perceive the test situation as painful and switch to behaviors indicative of having been through an aversive experience." In 2012, Rose and others reviewed this and further studies which concluded that pain had been found in fish. They concluded that the results from such research are due to poor design and misinterpretation, and that the researchers were unable to distinguish unconscious detection of injurious stimuli (nociception) from conscious pain.

Electroreception

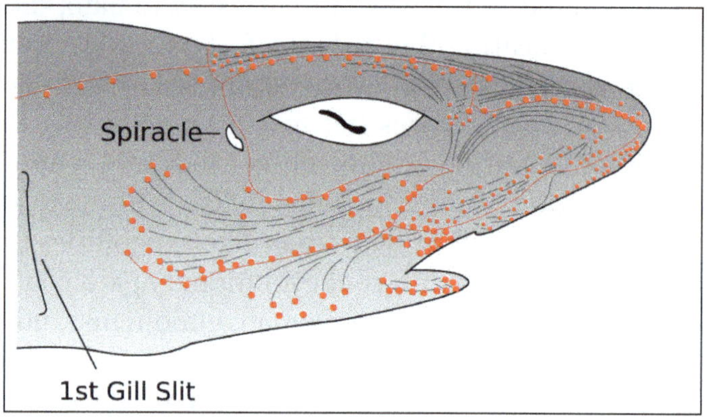

Electroreceptors (Ampullae of Lorenzini) and lateral line canals in the head of a shark.

Electroreception is the biological ability to perceive natural electrical stimuli. It has been observed almost exclusively in aquatic or amphibious animals, since salt-water is a much better conductor than air, the currently known exceptions being the platypus, echidnas, cockroaches and bees. Electroreception is used in electrolocation (detecting objects) and for electrocommunication.

Overview

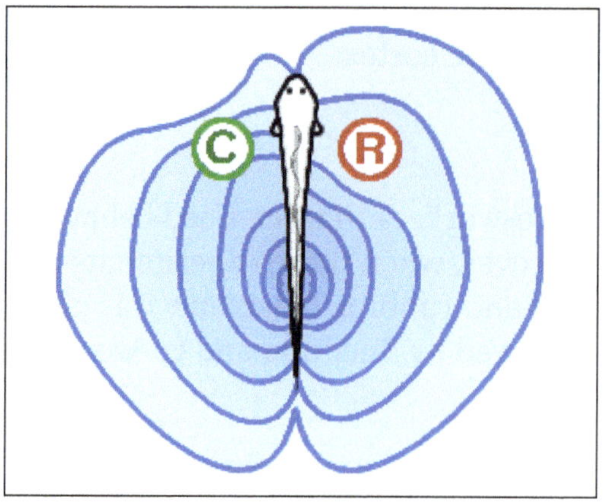

Active electrolocation. Conductive objects concentrate the field and resistive objects spread the field.

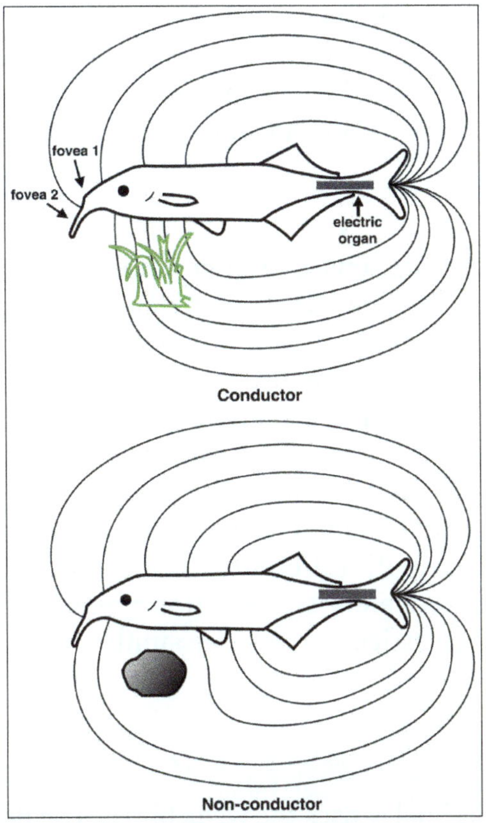

For the elephantfish (here *Gnathonemus*) the electric field emanates from an electric organ in the tail region (gray rectangle). It is sensed by the electroreceptive skin areas, using two electric pits (foveas) to actively search and inspect objects. Shown are the field distortions created by two different types of objects: a plant that conducts better than water, above (green) and a non-conducting stone, below (gray).

Until recently, electroreception was known only in vertebrates. Recent research has shown that bees can detect the presence and pattern of a static charge on flowers. Electroreception is found in lampreys, cartilaginous fishes (sharks, rays, chimaeras), lungfishes, bichirs, coelacanths, sturgeons, paddlefishes, catfishes, gymnotiformes, elephantfishes, monotremes, and at least one species of cetacean. The electroreceptor organs in all these groups are derived embryologically from a mechanoreceptor system. In fishes they are developed from the lateral lines. In most groups electroreception is *passive*, where it is used predominantly in predation. Two groups of teleost fishes are weakly electric and engage in *active* electroreception; the Neotropical knifefishes (Gymnotiformes) and the African elephantfishes (Notopteroidei). A rare terrestrial exception is the Western long-beaked echidna which has about 2,000 electroreceptors on its bill, compared to 40,000 for its semi-aquatic monotreme relative, the duck-billed platypus.

Electrolocation

Electroreceptive animals use this sense to locate objects around them. This is important in ecological niches where the animal cannot depend on vision: for example in caves, in murky water and at night. Many fish use electric fields to detect buried prey. Some shark embryos and pups "freeze" when they detect the characteristic electric signal of their predators. It has been proposed that sharks can use their acute electric sense to detect the earth's magnetic field by detecting the weak electric currents induced by their swimming or by the flow of ocean currents. The walking behaviour of cockroaches can be affected by the presence of a static electric field: they like to avoid the electric field. Cabbage loopers are also known to avoid electric fields.

Active Electrolocation

In active electrolocation, the animal senses its surrounding environment by generating electric fields and detecting distortions in these fields using electroreceptor organs. This electric field is generated by means of a specialised electric organ consisting of modified muscle or nerves. This field may be modulated so that its frequency and wave form are unique to the species and sometimes, the individual (see Jamming avoidance response). Animals that use active electroreception include the weakly electric fish, which either generate small electrical pulses (termed "pulse-type") or produce a quasi-sinusoidal discharge from the electric organ (termed "wave-type"). These fish create a potential which is usually smaller than one volt. Weakly electric fish can discriminate between objects with different resistance and capacitance values, which may help in identifying the object. Active electroreception typically has a range of about one body length, though objects with an electrical impedance similar to that of the surrounding water are nearly undetectable.

Passive Electrolocation

In passive electrolocation, the animal senses the weak bioelectric fields generated by other animals and uses it to locate them. These electric fields are generated by all animals due to the activity of their nerves and muscles. A second source of electric fields in fish is the ion pumps associated with osmoregulation at the gill membrane. This field is modulated by the opening and closing of the mouth and gill slits. Many fish that prey on electrogenic fish use the discharges of their prey to detect them. This has driven the prey to evolve more complex or higher frequency signals that are harder to detect.

Passive electroreception is carried out solely by ampullary electroreceptors in fish. It is tuned to low frequency signals (less than 1 Hz to tens of Hz).

Fish use passive electroreception to supplement or replace their other senses when detecting prey and predators. In sharks, sensing an electric dipole alone is sufficient to cause them to try to eat it.

Electrocommunication

Weakly electric fish can also communicate by modulating the electrical waveform they generate, an ability known as electrocommunication. They may use this for mate attraction and territorial displays. Some species of catfish use their electric discharges only in agonistic displays.

In one species of *Brachyhypopomus* (a genus of South American river fish belonging to the family Hypopomidae, commonly known as *bluntnose knifefishes*), the electric discharge pattern is similar to the low voltage electrolocative discharge of the electric eel. This is hypothesised to be a form of Batesian mimicry of the dangerous eel.

Sensory Mechanism

Active electroreception relies upon tuberous electroreceptors which are sensitive to high frequency (20-20,000 Hz) stimuli. These receptors have a loose plug of epithelial cells which capacitively couples the sensory receptor cells to the external environment. Passive electroreception however, relies upon ampullary receptors which are sensitive to low frequency stimuli (below 50 Hz). These receptors have a jelly-filled canal leading from the sensory receptors to the skin surface. Mormyrid electric fish from Africa use tuberous receptors known as Knollenorgans to sense electric communication signals.

Examples

Sharks and Rays

Sharks and rays (members of the subclass *Elasmobranchii*), such as the lemon shark, rely heavily on electrolocation in the final stages of their attacks, as can be demonstrated by the robust feeding response elicited by electric fields similar to those of their prey. Sharks are the most electrically sensitive animals known, responding to DC fields as low as 5 nV/cm.

The electric field sensors of sharks are called the ampullae of Lorenzini. They consist of electroreceptor cells connected to the seawater by pores on their snouts and other zones of the head. A problem with the early submarine telegraph cables was the damage caused by sharks who sensed the electric fields produced by these cables. It is possible that sharks may use Earth's magnetic field to navigate the oceans using this sense.

Bony Fish

The electric eel, besides its ability to generate high voltage electric shocks, uses lower voltage pulses for navigation and prey detection in its turbid habitat. This ability is shared with other gymnotiformes.

Monotremes

The monotremes are the only group of land mammals known to have evolved electroreception. While the electroreceptors in fish and amphibians evolved from mechanosensory lateral line organs, those of monotremes are based on cutaneous glands innervated by trigeminal nerves. The electroreceptors of monotremes consist of free nerve endings located in the mucous glands of the snout. Among the monotremes, the platypus (*Ornithorhynchus anatinus*) has the most acute electric sense. The platypus has almost 40,000 electroreceptors arranged in a series of stripes along the bill, which probably aids the localisation of prey. The platypus electroreceptive system is highly directional, with the axis of greatest sensitivity pointing outwards and downwards. By making short-latency head movements called "saccades" when swimming, platypuses constantly expose the most sensitive part of their bill to the stimulus to localise prey as accurately as possible. The platypus appears to use electroreception along with pressure sensors to determine the distance to prey from the delay between the arrival of electrical signals and pressure changes in the water.

The platypus also uses electroreception.

The electroreceptive capabilities of the two species of echidna (which are terrestrial) are much more simple. Long-beaked echidnas (genus *Zaglossus*) possess only 2,000 receptors and short-beaked echidnas (*Tachyglossus aculeatus*) merely 400 that are concentrated in the tip of the snout. This difference can be attributed to their habitat and feeding methods. Western long-beaked echidnas live in wet tropical forests where they feed on earthworms in damp leaf litter, so their habitat is probably favourable to the reception of electrical signals. Contrary to this is the varied but generally more arid habitat of their short-beaked relative which feeds primarily on termites and ants in nests; the humidity in these nests presumably allows electroreception to be used in hunting for buried prey, particularly after rains. Experiments have shown that echidnas can be trained to respond to weak electric fields in water and moist soil. The electric sense of the echidna is hypothesised to be an evolutionary remnant from a platypus-like ancestor.

Dolphins

Dolphins have evolved electroreception in structures different from those of fish, amphibians and monotremes. The hairless vibrissal crypts on the rostrum of the Guiana dolphin (*Sotalia guianensis*), originally associated with mammalian whiskers, are capable of electroreception as low as 4.8 μV/cm, sufficient to detect small fish. This is comparable to the sensitivity of electroreceptors in the platypus. To date (June 2013), these cells have been described from only a single dolphin specimen.

Bees

Bees collect a positive static charge while flying through the air (see Atmospheric electricity). When a bee visits a flower, the charge deposited on the flower takes a while to leak away into the ground. Bees can detect both the presence and the pattern of electric fields on flowers, and use this information to know if a flower has been recently visited by another bee and is therefore likely to have a reduced concentration of nectar. The mechanism of electric field reception in animals living in the air like bees is based on mechano- reception, not electroreception. Bees receive the electric field changes via the Johnston's organs in their antennae and possibly other mechano-receptors. They distinguish different temporal patterns and learn them. During the waggle dance, honeybees appear to use the electric field emanating from the dancing bee for distance communication.

Effects on Wildlife

It has been claimed that the electromagnetic fields generated by pylons and masts have adverse effects on wildlife; a list of 153 references to this has been published.

Fish Intelligence

The elephantnose fish has the highest brain-to-body oxygen consumption ratio of all known vertebrates.

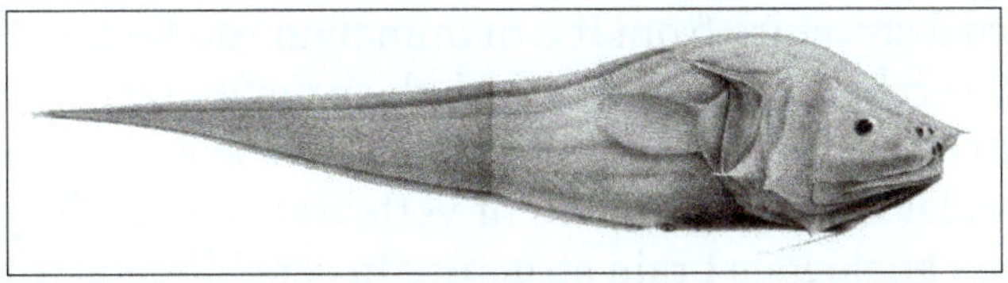

The bony-eared assfish has the smallest brain-to-body weight ratio of all known vertebrates.

Fish intelligence is "...the resultant of the process of acquiring, storing in memory, retrieving, combining, comparing, and using in new contexts information and conceptual skills" as it applies to fish.

According to Culum Brown from Macquarie University, "Fish are more intelligent than they appear. In many areas, such as memory, their cognitive powers match or exceed those of 'higher' vertebrates including non-human primates."

Fish hold records for the relative brain weights of vertebrates. Most vertebrate species have similar brain-to-body mass ratios. The deep sea bathypelagic bony-eared assfish, has the smallest ratio of all known vertebrates. At the other extreme, the electrogenic elephantnose fish, an African freshwater fish, has one of the largest brain-to-body weight ratios of all known vertebrates (slightly higher than humans) and the highest brain-to-body oxygen consumption ratio of all known vertebrates (three times that for humans).

Brain

Fish typically have quite small brains relative to body size compared with other vertebrates, typically one-fifteenth the brain mass of a similarly sized bird or mammal. However, some fish have relatively large brains, most notably mormyrids and sharks, which have brains about as massive relative to body weight as birds and marsupials.

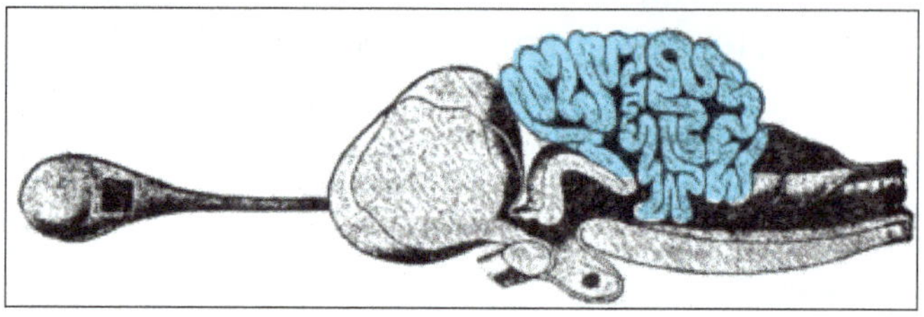

Cross-section of the brain of a porbeagle shark, with the cerebellum highlighted.

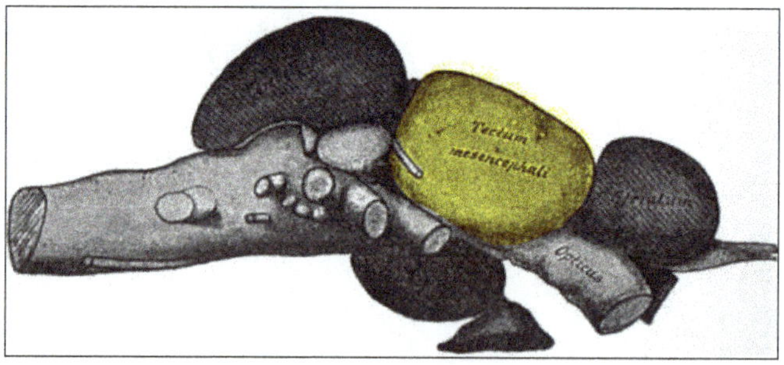

The brain of a cod.

The cerebellum of cartilaginous and bony fishes is large and complex. In at least one important respect, it differs in internal structure from the mammalian cerebellum: The fish cerebellum does not contain discrete deep cerebellar nuclei. Instead, the primary targets of Purkinje cells are a distinct type of cell distributed across the cerebellar cortex, a type not seen in mammals. The circuits in the cerebellum are similar across all classes of vertebrates, including fish, reptiles, birds, and mammals. There is also an analogous brain structure in cephalopods with well-developed brains, such as octopuses. This has been taken as evidence that the cerebellum performs functions important to all animal species with a brain.

In mormyrid fish (a family of weakly electrosensitive freshwater fish), the cerebellum is considerably larger than the rest of the brain put together. The largest part of it is a special structure called the *valvula*, which has an unusually regular architecture and receives much of its input from the electrosensory system.

Memory

Red Sea clownfish can recognize their mate after 30 days separation.

Individual carp captured by anglers have been shown to become less catchable thereafter. This suggests that fish use their memory of negative experiences to associate capture with stress and therefore become less easy to catch. This type of associative learning has also been shown in paradise fish (*Macropodus opercularis*) which avoid places where they have experienced a single attack by a predator and continue to avoid for many months.

Some fish species can exhibit long-term memory. Anecdotally, channel catfish (*Ictalurus punctatus*) can remember the human voice call announcing food five years after last hearing that call. Goldfish remember the colour of a tube dispensing food one year after the last tube presentation. Sockeye salmon still react to a light signal that precedes food arrival up to eight months since the last reinforcement. Some common rudd and European chub could remember the person who trained them to feed from the hand, even after a 6-month break. Crimson-spotted rainbowfish can learn how to escape from a trawl by swimming through a small hole in the center and they remember this technique 11 months later. Rainbow trout can be trained to press a bar to get food, and they remember this three months after last seeing the bar. Red Sea clownfish can recognize their mate 30 days after it was experimentally removed from the home anemone.

Several fish species are capable of learning complex spatial relationships and forming cognitive maps. They can orient themselves using multiple landmarks or symbols and they are able to integrate experiences which enable them to generate appropriate avoidance responses.

Tool Use

Tool use is sometimes considered as an indication of intelligence in animals. There are few examples of tool use in fishes, perhaps because they have only their mouth in which to hold objects.

Several species of wrasse hold bivalves (scallops, clams and urchins) in their mouth and smash them against the surface of a rock (an "anvil") to break them up. This behaviour in an orange-dotted tuskfish (*Choerodon anchorago*) has been filmed; the fish fans sand to unearth the bivalve, takes it into its mouth, swims several metres to a rock which it uses as an anvil by smashing the mollusc apart with sideward thrashes of the head.

Archerfish (family Toxotidae) squirt jets of water at insects on plants above the surface to knock them into the water; they can adjust the size of the squirts to the size of the insect prey and learn to shoot at moving targets.

Whitetail damselfish clean the rock face where they intend to lay eggs by sucking up and blowing sand grains onto the surface. Triggerfish blow water at sea urchins to turn them over, thereby exposing their more vulnerable underside. Banded acaras (*Bujurquina vittata*) lay their eggs on a loose leaf and carry the leaf away when a predator approaches.

In one laboratory study, Atlantic cod (*Gadus morhua*) given access to an operant feeding machine learned to pull a string to get food. However, the researchers had also tagged the fish by threading a bead in front of their dorsal fin. Some fish snagged the string with their bead, resulting in food delivery. These fish eventually learned to swim in a particular way to repeatedly make the bead snag the string and get food. Because the fish used an object external to their body in a goal-oriented way, this satisfies some definitions of tool use.

Construction

As for tool use, construction behaviour may be mostly innate. Yet it can be sophisticated, and the fact that fish can make judicious repairs to their creation suggests intelligence. Construction methods in fishes can be divided into three categories: excavations, pile-ups, and gluing.

Excavations may be simple depressions dug up in the substrate, such as the nests of bowfin, smallmouth bass, and Pacific salmon, but it can also consist of fairly large burrows used for shelter and for nesting. Burrowing species include the mudskippers, the red band-fish *Cepola rubescens* (burrows up to 1 m deep, often with a side branch), the yellowhead jawfish *Opistognathus aurifrons* (chambers up to 22 cm deep, lined with coral fragments to solidify it), the convict blenny *Pholidichthys leucotaenia* whose burrow is a maze of tunnels and chambers thought to be as much as 6 m long, and the Nicaragua cichlid, *Hypsophrys nicaraguensis*, who drills a tunnel by spinning inside of it. In the case of the mudskippers, the burrows are shaped like a J and can be as much as 2 m deep. Two species, the giant mudskipper *Periophthalmodon schlosseri* and the walking goby *Scartelaos histophorus*, build a special chamber at the bottom of their burrows into which they carry mouthfuls of air. Once released the air accumulates at the top of the chamber and forms a reserve from which the fish can breathe – like all amphibious fishes, mudskippers are good air breathers. If researchers experimentally extract air from the special chambers, the fish diligently replenish it. The significance of this behaviour stems from the facts that at high tide, when water covers the mudflats, the fish stay in their burrow to avoid predators, and water inside the confined burrow is often poorly oxygenated. At such times these air-breathing fishes can tap into the air reserve of their special chambers .

Mounds are easy to build, but can be quite extensive. In North American streams, the male cutlip minnow *Exoglossum maxillingua*, 90–115 mm long (3.5-4.5 in), assembles mounds that are 75–150 mm high (3–6 in), 30–45 cm in diameter (12–18 in), made up of more than 300 pebbles 13–19 mm in diameter (a quarter to half an inch). The fish carry these pebbles one by one in their mouths, sometimes stealing some from the mounds of other males. The females deposit their eggs on the upstream slope of the mounds, and the males cover these eggs with more pebbles. Males of the hornyhead chub *Nocomis biguttatus*, 90 mm long (3.5 in), and of the river chub *Nocomis micropogon*, 100 mm long (4 in), also build mounds during the reproductive season. They start by clearing a slight depression in the substrate, which they overfill with up to 10,000 pebbles until the mounds are 60–90 cm (2–3 ft) long (in the direction of the water current), 30–90 cm wide (1–3 ft), and 5–15 cm high (2–6 in). Females lay their eggs among those pebbles. The stone accumulation is free of sand and it exposes the eggs to a good water current that supplies oxygen. Males of many mouthbrooding cichld species in Lake Malawi and Lake Tanganyika build sand cones that are flattened or crater-shaped at the top. Some of these mounds can be 3 m in diameter and 40 cm high. The mounds serve to impress females or to allow species recognition during courtship.

Male pufferfish, *Torquigener* sp., also build sand mounds to attract females. The mounds, up to 2 m in diameter, are intricate with radiating ridges and valleys.

Several species build up mounds of coral pieces either to protect the entrance to their burrows, as in tilefishes and gobies of the genus Valenciennea, or to protect the patch of sand in which they will bury themselves for the night, as in the Jordan's tuskfish *Choerodon jordani* and the rockmover wrasse *Novaculichthys taeniourus*.

Male sticklebacks are well known for their habit of building an enclosed nest made of pieces of vegetation glued together with secretions from their kidneys. Foam nests, made up of air bubbles glued together with mucus from the mouth, are also well known in gouramis and armoured catfish.

Social Intelligence

Fish can remember the attributes of other individuals, such as their competitive ability or past behavior, and modify their own behavior accordingly. For example, they can remember the identity of individuals to whom they have lost in a fight, and avoid these individuals in the future; or they can recognize territorial neighbors and show less aggression towards them as compared to strangers. They can recognize individuals in whose company they obtained less food in the past and preferentially associate with new partners in the future.

Fish can seem mindful of which individuals have watched them in the past. In an experiment with Siamese fighting fish, two males were made to fight each other while being watched by a female, whom the males could also see. The winner and the loser of the fight were then, separately, given a choice between spending time next to the watching female or to a new female. The winner courted both females equally, but the loser spent more time next to the new female, avoiding the watcher female. In this species, females prefer males they have seen win a fight over males they have seen losing, and it therefore makes sense for a male to prefer a female that has never seen him as opposed to a female that has seen him lose.

Knowing that if A>B and B>C, then A>C, is another type of evidence for intelligence, and it can be applied in the context of dominance hierarchies. In a study with the cichlid *Astatotilapia burtoni*, eight observer fish could watch individual A beat individual B, then B over C, C over D, and D over E. The observer fish were then given a choice of associating with either B or D (both of which they had seen win once and lose once). All eight observer fish spent more time next to D. Fish in this species prefer to associate with more subordinate individuals, so the preference for D showed that the observers had worked out that B was superior to C, and C to D, and therefore D was subordinate to B.

Deception

There are several examples of fish being deceptive, indicating they are capable of a theory of mind.

Distraction Display

In the threespine stickleback (*Gasterosteus aculeatus*), males sometimes see their nest full of eggs fall prey to groups of marauding females; some males, when they see a group of females approaching, swim away from their nest and start poking their snout into the substrate, as would a female raiding a nest. This distraction display commonly fools the females into behaving as if a nest has been discovered there and they rush to that site, leaving the male's real nest alone. Bowfin (*Amia calva*) males caring for their free-swimming fry exhibit a related distraction display when a potential fry-predator approaches; they move away and thrash about as if injured, drawing the predator's attention toward himself.

Adult male bowfins distract potential predators away from their fry by thrashing as if injured.

False Courtship Behaviour

In the Malili Lakes of Sulawesi, Indonesia, one species of sailfin silverside (*Telmatherina sarasinorum*) is an egg predator. They often follow courting pairs of the closely related species *T. antoniae*. When those pairs lay eggs, *T. sarasinorum* darts in and eats the eggs. On four different occasions in the field (out of 136 observation bouts in total), a male *T. sarasinorum* who was following a pair of courting *T. antoniae* eventually chased off the male *T. antoniae* and took his place, courting the heterospecific female. That female released eggs, at which point the male darted to the eggs and ate them.

Death Feigning

Death feigning as a way to attract prey is another form of deception. In Lake Malawi, the predatory cichlid *Nimbochromis livingstonii* have been seen first remaining stationary with their abdomen on or near sand and that then dropping onto their sides. In a variant behaviour, some *N. livingstonii* fell through the water column and landed onto their side. The fish then remained immobile for several minutes. Their colour pattern was blotchy and suggested a rotting carcass. Small inquisitive cichlids of other species often came near and they were suddenly attacked by the predator. About a third of the death-feigning performances led to an attack, and about one-sixth of the attacks were successful. Another African cichlid, *Lamprologus lemairii*, from Lake Tanganyika, has been reported to do the same thing. A South American cichlid, the yellowjacket cichlid *Parachromis friedrichsthalii*, also uses death feigning. They turn over onto their sides at the bottom of the sinkholes they inhabit and remain immobile for as long as 15 minutes, during which they attack the small mollies that come too close to them. The comb grouper *Mycteroperca acutirostris* may also be an actor, though in this case the behaviour should be called dying or illness feigning, rather than death feigning, because while lying on its side the fish occasionally undulates its body. In 1999, off the coast of southeastern Brazil, one juvenile comb grouper was observed using this tactic to catch five small prey in 15 minutes.

Cooperation

Cooperative foraging reflects some mental flexibility and planning, and could therefore be interpreted as intelligence. There are a few examples in fishes.

Yellowtail amberjack can form packs of 7-15 individuals that maneuver in U-shaped formations to cut away the tail end of prey shoals (jack mackerels or Cortez grunts) and herd the downsized shoal next to seawalls where they proceed to capture the prey.

In the coral reefs of the Red Sea, roving coralgrouper that have spotted a small prey fish hiding in a crevice sometimes visit the sleeping hole of a giant moray and shake their head at the moray, and this seems to be an invitation to group hunting as the moray often swims away with the grouper, is led to the crevice where the prey hides, and proceeds to probe that crevice (which is too small to let the grouper in), either catching the prey by itself or flushing it into the open where the grouper grabs it. The closely related coral trout also enrolls the help of moray eels in this way, and they only do so when the prey they seek is hidden in crevices, where only the eel can flush them. They also quickly learn to invite preferentially those individual eels that collaborate most often.

Similarly, zebra lionfish that detect the presence of small prey fishes flare up their fins as an invitation to other zebra lionfish, or even to another species of lionfish (*Pterois antennata*), to join them in better cornering the prey and taking turns at striking the prey so that every individual hunter ends up with similar capture rates.

Numeracy

Mosquitofish (*Gambusia holbrooki*) can distinguish between doors marked with either two or three geometric symbols, only one of which allows the fish to rejoin its shoalmates. This can be achieved when the two symbols have the same total surface area, density and brightness as the three symbols. Further studies show this discrimination extends to 4 vs 8, 15 vs 30, 100 vs 200, 7 vs 14, and 8 vs 12 symbols, again controlling for non-numerical factors.

Many studies have shown that when given a choice, shoaling fish prefer to join the largest of two shoals. It has been argued that several aspects of such choice reflect an ability by fish to distinguish between numerical quantities.

Social Learning

Fish can learn how to perform a behavior simply by watching other individuals in action. This is variously called observational learning, cultural transmission, or social learning. For example, fish can learn a particular route after following an experienced leader a few times. One study trained guppies to swim through a hole marked in red while ignoring another one marked in green in order to get food on the other side of a partition; when these experienced fish ("demonstrators") were joined by a naive one (an "observer"), the observer followed the demonstrators through the red hole, and kept the habit once the demonstrators were removed, even when the green hole now allowed food access. In the wild, juvenile French grunt follow traditional migration routes, up to 1 km long, between their daytime resting sites and their nighttime foraging areas on coral reefs; if groups of 10-20 individuals are marked and then transplanted to new populations, they follow the residents along what is for them – the transplants – a new migration route, and if the residents are then removed two days later, the transplanted grunts continue to use the new route, as well as the resting and foraging sites at both ends.

Through cultural transmission, fishes could also learn where good food spots are. Ninespine stickleback, when given a choice between two food patches they have watched for a while, prefer the patch over which more fish have been seen foraging, or over which fish were seen feeding more intensively. Similarly, in a field experiment where Trinidadian guppies were given a choice between two distinctly marked feeders in their home rivers, the subjects chose the feeder where other guppies were already present, and in subsequent tests when both feeders were deserted, the subjects remembered the previously popular feeder and chose it.

Through social learning, fishes might learn not only where to get food, but also what to get and how to get it. Hatchery-raised salmon can be taught to quickly accept novel, live prey items similar to those they will encounter once they will be released in the wild, simply by watching an experienced salmon take such prey. The same is true of young perch. In the laboratory, juvenile European seabass can learn to push a lever in order to obtain food just by watching experienced individuals use the lever.

Fishes can also learn from others the identity of predatory species. Fathead minnows, for example, can learn the smell of a predatory pike just by being simultaneously exposed to that smell and the sight of experienced minnows reacting with fear, and brook stickleback can learn the visual identity of a predator by watching the fright reaction of experienced fathead minnows. Fish can also learn to recognize the odor of dangerous sites when they are simultaneously exposed to it and to other fish that suddenly show a fright reaction. Hatchery-raised salmon can learn the smell of a predator by being simultaneously exposed to it and to the alarm substance released by injured salmon.

Latent Learning

Latent learning is a form of learning that is not immediately expressed in an overt response; it occurs without any obvious reinforcement of the behaviour or associations that are learned. One example in fish comes from research with male three spot gouramis (*Trichopodus trichopterus*). This species quickly form dominance hierarchies. To appease dominants, subordinates adopt a typical body posture angled at 15-60° to the horizontal, all fins folded and pale body colors. Individuals trained to associate a light-stimulus with the imminent arrival of food exhibit this associative learning by approaching the surface where the food is normally dropped immediately the light-stimulus is presented. However, if a subordinate is placed in a tank with a dominant individual and the light-stimulus is presented, the subordinate immediately assumes the submissive posture rather than approaching the surface. The subordinate has predicted that going to the surface to get food would place it in competition with the dominant, and to avoid potential aggression, it immediately attempts to appease the dominant.

Cleaner Fish

The bluestreak cleaner wrasse (*Labroides dimidiatus*) performs a service for "client" fishes (belonging to other species) by removing and eating their ectoparasites. Clients can invite a cleaning session by adopting a typical posture or simply by remaining immobile near a wrasse's cleaning station. They can even form queues while doing so. But cleaning sessions do not always end up well, because wrasses (or wrasse-mimicking parasitic sabre-toothed blennies) may cheat and eat the nutritious body mucus of their clients, rather than just the ectoparasites, something that makes the client jolt and sometimes flee. This system has been the subject of extensive observations which have suggested cognitive abilities on the part of the cleaner wrasses and their clients. For example, clients refrain from soliciting a cleaning session if they have witnessed the cleaning session of the previous client ending badly. Cleaners give the impression of trying to maintain a good reputation, because they cheat less when they see a big audience (a long queue of clients) watching. Cleaners sometimes work as male-female teams, and when the smaller female cheats and bites the client, the larger male chases her off, as if to punish her for having tarnished their reputation.

Bluestreak cleaner wrasse (bottom) with a client fish.

Play

Play behaviour is often considered a correlate of intelligence. One possible example in fish is provided by the electrolocating Peters' elephantnose fish (mentioned above as having one of the largest brain-to-body weight ratios of all known vertebrates). One captive individual was observed carrying a small ball of aluminum foil (a good conductor of electricity) to the outflow tube of the aquarium filter, letting the current push the ball away before chasing after it and repeating the behaviour.

Food Stocking

Food stocking can be viewed potentially as an animal planning for the future. One example of short-term stocking involves climbing perch (*Anabas testudineus*). Individuals were kept singly in aquaria and fed with pellets dropped at the surface. When the pellets were dropped one after the other at 1-s intervals, the fish took them as they reached the surface and stocked them inside the mouth. On average, the fish placed 7 pellets in their mouth before moving away to consume them. When starved for 24-h before the feeding test, they doubled the number of pellets stocked (14 on average); the underside of their heads bulged under the load. The behaviour may be an indication that competition for food is normally severe in this species and that any adaptation to secure food would be beneficial.

References

- Reimchen, T E; Temple, N F (2004). "Hydrodynamic and phylogenetic aspects of the adipose fin in fishes". Canadian Journal of Zoology. 82 (6): 910–916. doi:10.1139/Z04-069
- Kapoor BG and Khanna B (2004) Ichthyology Handbook pp. 497–498, Springer Science & Business Media. ISBN 9783540428541

- Dement J Species Spotlight: Atlantic Sailfish (Istiophorus albicans) Archived December 17, 2010, at the Wayback Machine. littoralsociety.org. Retrieved 1 April 2012
- Sharpe, P. T. (2001). "Fish scale development: Hair today, teeth and scales yesterday?". Current Biology. 11 (18): R751–R752. PMID 11566120. doi:10.1016/S0960-9822(01)00438-9
- Nauen, JC; Lauder, GV (2000). "Locomotion in scombrid fishes: morphology and kinematics of the finlets of the Chub mackerel Scomber japonicus" (PDF). Journal of Experimental Biology. 203: 2247–59
- Meyer CG, Holland KN, Papastamatiou YP (2005). "Sharks can detect changes in the geomagnetic field". Journal of the Royal Society, Interface. 2 (2): 129–30. PMC 1578252. PMID 16849172. doi:10.1098/rsif.2004.0021
- "Modifications of the Digestive Tract for Holding Air in Loricariid and Scoloplacid Catfishes" (PDF). Copeia (3): 663–675. 1998. doi:10.2307/1447796. Retrieved 25 June 2009
- Fowler, Sarah; Cavanagh, Rachel. "Sharks, Rays, and Chimaeras: The Status of Chondrichthyan Fishes" (PDF). The World Conservation Union. Retrieved 3 May 2017
- Reddon, AR; O'Connor, CM; Marsh-Rollo, SE; Balcshine, S (2012). "Effects of isotocin on social responses in a cooperatively breeding fish". Animal Behaviour. 84 (4): 753–760. doi:10.1016/j.anbehav.2012.07.021
- Märss, T. (2006). "Exoskeletal ultrasculpture of early vertebrates". Journal of Vertebrate Paleontology. 26 (2): 235–252. doi:10.1671/0272-4634(2006)26[235:EUOEV]2.0.CO;2

Chapter 4

Reproduction in Fishes

Fishes are unisexual and reproduce sexually. Most fishes are oviparous and the fertilization is external. Many adult fishes of the same species come together in a group and release their gametes into the water at the same time. This process is known as spawning. The topics elaborated in this chapter will help in gaining a better perspective about the process reproduction in fishes.

Fish Reproduction

Fish reproductive organs include testes and ovaries. In most species, gonads are paired organs of similar size, which can be partially or totally fused. There may also be a range of secondary organs that increase reproductive fitness. The genital papilla is a small, fleshy tube behind the anus in some fishes, from which the sperm or eggs are released; the sex of a fish often can be determined by the shape of its papilla.

Anatomy

Testes

Most male fish have two testes of similar size. In the case of sharks, the testes on the right side is usually larger. The primitive jawless fish have only a single testis, located in the midline of the body, although even this forms from the fusion of paired structures in the embryo.

Under a tough membranous shell, the tunica albuginea, the testis of some teleost fish, contains very fine coiled tubes called seminiferous tubules. The tubules are lined with a layer of cells (germ cells) that from puberty into old age, develop into sperm cells (also known as spermatozoa or male gametes). The developing sperm travel through the seminiferous tubules to the rete testis located in the mediastinum testis, to the efferent ducts, and then to the epididymis where newly created sperm cells mature. The sperm move into the vas deferens, and are eventually expelled through the urethra and out of the urethral orifice through muscular contractions.

However, most fish do not possess seminiferous tubules. Instead, the sperm are produced in spherical structures called *sperm ampullae*. These are seasonal structures, releasing their contents during the breeding season, and then being reabsorbed by the body. Before the next breeding season, new sperm ampullae begin to form and ripen. The ampullae are otherwise essentially identical to the seminiferous tubules in higher vertebrates, including the same range of cell types.

In terms of spermatogonia distribution, the structure of teleosts testes has two types: in the most common, spermatogonia occur all along the seminiferous tubules, while in Atherinomorph fish they are confined to the distal portion of these structures. Fish can present cystic or semi-cystic spermatogenesis in relation to the release phase of germ cells in cysts to the seminiferous tubules lumen.

Ovaries

Many of the features found in ovaries are common to all vertebrates, including the presence of follicular cells and tunica albuginea There may be hundreds or even millions of fertile eggs present in the ovary of a fish at any given time. Fresh eggs may be developing from the germinal epithelium throughout life. Corpora lutea are found only in mammals, and in some elasmobranch fish; in other species, the remnants of the follicle are quickly resorbed by the ovary. The ovary of teleosts is often contains a hollow, lymph-filled space which opens into the oviduct, and into which the eggs are shed. Most normal female fish have two ovaries. In some elasmobranchs, only the right ovary develops fully. In the primitive jawless fish, and some teleosts, there is only one ovary, formed by the fusion of the paired organs in the embryo.

Fish ovaries may be of three types: gymnovarian, secondary gymnovarian or cystovarian. In the first type, the oocytes are released directly into the coelomic cavity and then enter the ostium, then through the oviduct and are eliminated. Secondary gymnovarian ovaries shed ova into the coelom from which they go directly into the oviduct. In the third type, the oocytes are conveyed to the exterior through the oviduct. Gymnovaries are the primitive condition found in lungfish, sturgeon, and bowfin. Cystovaries characterize most teleosts, where the ovary lumen has continuity with the oviduct. Secondary gymnovaries are found in salmonids and a few other teleosts.

Eggs

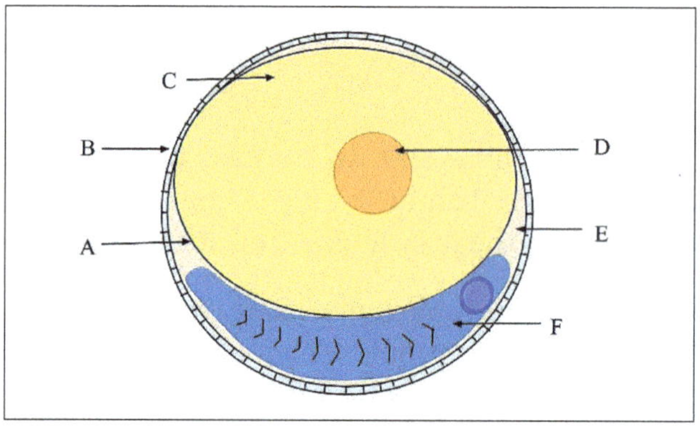

Diagram of a fish egg: A. vitelline membrane B. chorion C. yolk D. oil globule E. perivitelline space F. embryo.

The eggs of fish and amphibians are jellylike. Cartilagenous fish (sharks, skates, rays, chimaeras) eggs are fertilized internally and exhibit a wide variety of both internal and external embryonic development. Most fish species spawn eggs that are fertilized externally, typically with the male inseminating the eggs after the female lays them. These eggs do not have a shell and would dry out in the air. Even air-breathing amphibians lay their eggs in water, or in protective foam as with the Coast foam-nest treefrog, *Chiromantis xerampelina.*

Intromittent Organs

Male cartilaginous fishes (sharks and rays), as well as the males of some live-bearing ray finned fishes, have fins that have been modified to function as intromittent organs, reproductive appendages which allow internal fertilization. In ray finned fish they are called *gonopodiums* or *andropodiums*, and in cartilaginous fish they are called *claspers*.

This male mosquitofish has a gonopodium, an anal fin which functions as an intromittent organ.

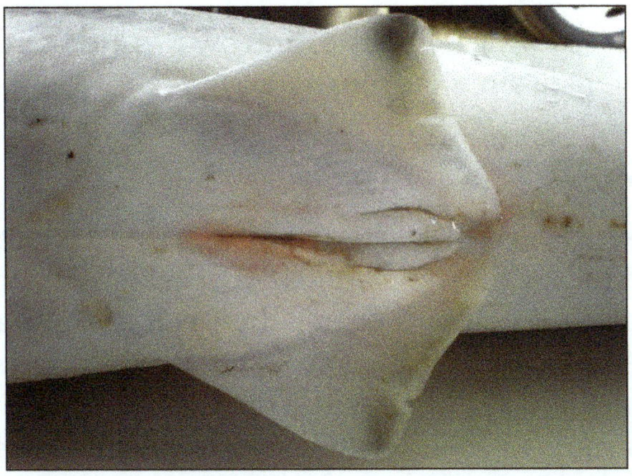

This young male spinner shark has claspers, a modification to the pelvic fins which also function as intromittent organs.

Gonopodia are found on the males of some species in the Anablepidae and Poeciliidae families. They are anal fins that have been modified to function as movable intromittent organs and are used to impregnate females with milt during mating. The third, fourth and fifth rays of the male's anal fin are formed into a tube-like structure in which the sperm of the fish is ejected. When ready for mating, the gonopodium becomes erect and points forward towards the female. The male shortly inserts the organ into the sex opening of the female, with hook-like adaptations that allow the fish to grip onto the female to ensure impregnation. If a female remains stationary and her partner contacts her vent with his gonopodium, she is fertilized. The sperm is preserved in the female's oviduct. This allows females to fertilize themselves at any time without further assistance from males. In some species, the gonopodium may be half the total body length. Occasionally the fin is too long to be used, as in the "lyretail" breeds of *Xiphophorus helleri*. Hormone treated females may develop gonopodia. These are useless for breeding.

Similar organs with similar characteristics are found in other fishes, for example the *andropodium* in the *Hemirhamphodon* or in the Goodeidae.

Claspers are found on the males of cartilaginous fishes. They are the posterior part of the pelvic fins that have also been modified to function as intromittent organs, and are used to channel semen into the female's cloaca during copulation. The act of mating in sharks usually includes raising one of the claspers to allow water into a siphon through a specific orifice. The clasper is then inserted into the cloaca, where it opens like an umbrella to anchor its position. The siphon then begins to contract expelling water and sperm.

Physiology

Oogonia development in teleosts fish varies according to the group, and the determination of oogenesis dynamics allows the understanding of maturation and fertilisation processes. Changes in the nucleus, ooplasm, and the surrounding layers characterize the oocyte maturation process.

Postovulatory follicles are structures formed after oocyte release; they do not have endocrine function, present a wide irregular lumen, and are rapidly reabsorbed in a process involving the apoptosis of follicular cells. A degenerative process called follicular atresia reabsorbs vitellogenic oocytes not spawned. This process can also occur, but less frequently, in oocytes in other development stages.

Some fish are hermaphrodites, having both testes and ovaries either at different phases in their life cycle or, as in hamlets, have them simultaneously.

Reproductive Strategies

In fish, fertilisation of eggs can be either external or internal. In many species of fish, fins have been modified to allow Internal fertilisation. Similarly, development of the embryo can be external or internal, although some species show a change between the two at various stages of embryo development. Thierry Lodé described reproductive strategies in terms of the development of the zygote and the interrelationship with the parents; there are five classifications - ovuliparity, oviparity, ovo-viviparity, histotrophic viviparity and hemotrophic viviparity.

Ovuliparity

Ovuliparity means the female lays unfertilised eggs (ova), which must then be externally fertilised. Examples of ovuliparous fish include salmon, goldfish, cichlids, tuna and eels. In the majority of these species, fertilisation takes place outside the mother's body, with the male and female fish shedding their gametes into the surrounding water.

Oviparity

Oviparity is where fertilisation occurs internally and so the female sheds zygotes (or newly developing embryos) into the water, often with important outer tissues added. Over 97% of all known fish are oviparous, In oviparous fish, internal fertilisation requires the male to use some sort of intromittent organ to deliver sperm into the genital opening of the female. Examples include the oviparous sharks, such as the horn shark, and oviparous rays, such as skates. In these cases, the male is equipped with a pair of modified pelvic fins known as claspers.

Marine fish can produce high numbers of eggs which are often released into the open water column. The eggs have an average diameter of 1 millimetre (0.039 in). The eggs are generally surrounded by the extraembryonic membranes but do not develop a shell, hard or soft, around these membranes. Some fish have thick, leathery coats, especially if they must withstand physical force or desiccation. These type of eggs can also be very small and fragile.

The newly hatched young of oviparous fish are called larvae. They are usually poorly formed, carry a large yolk sac (for nourishment) and are very different in appearance from juvenile and adult specimens. The larval period in oviparous fish is relatively short (usually only several weeks), and larvae rapidly grow and change appearance and structure (a process termed metamorphosis) to become juveniles. During this transition larvae must switch from their yolk sac to feeding on zooplankton prey, a process which depends on typically inadequate zooplankton density, starving many larvae.

Ovoviviparity

In ovoviviparous fish the eggs develop inside the mother's body after internal fertilisation but receive little or no nourishment directly from the mother, depending instead on a food reserve inside the egg, the yolk. Each embryo develops in its own egg. Familiar examples of ovoviviparous fish include guppies, angel sharks, and coelacanths.

Viviparity

There are two types of viviparity, differentiated by how the offspring gain their nutrients.

- Histotrophic (tissue eating) viviparity means embryos develop in the female's oviducts but obtain nutrients by consuming other tissues, such as ova (oophagy) or zygotes. This has been observed primarily among sharks such as the shortfin mako and porbeagle, but is known for a few bony fish as well such as the halfbeak *Nomorhamphus ebrardtii*. An unusual mode of vivipary is adelphophagy or intrauterine cannibalism, in which the largest embryos eat weaker, smaller unborn siblings. This is most commonly found among sharks such as the grey nurse shark, but has also been reported for *Nomorhamphus ebrardtii*.
- Hemotrophic (blood eating) viviparity means embryos develop in the female's (or male's) oviduct and nutrients are provided directly by the parent, typically via a structure similar to, or analogous to the placenta seen in mammals. Examples of hemotrophic fish include the surfperches, splitfins, lemon shark, seahorses and pipefish.

Aquarists commonly refer to ovoviviparous and viviparous fish as livebearers.

Hermaphroditism

Female groupers change their sex to male if no male is available.

An anemone fish couple guarding their anemone. If the female dies, a juvenile male moves in, and the resident male changes sex.

Parthenogenesis was first described among vertebrates in the Amazon molly.

Hermaphroditism occurs when a given individual in a species possesses both male and female reproductive organs, or can alternate between possessing first one, and then the other. Hermaphroditism is common in invertebrates but rare in vertebrates. It can be contrasted with gonochorism, where each individual in a species is either male or female, and remains that way throughout their lives. Most fish are gonochorists, but hermaphroditism is known to occur in 14 families of teleost fishes.

Usually hermaphrodites are *sequential*, meaning they can switch sex, usually from female to male (protogyny). This can happen if a dominant male is removed from a group of females. The largest female in the harem can switch sex over a few days and replace the dominant male. This is found amongst coral reef fishes such as groupers, parrotfishes and wrasses. It is less common for a male to switch to a female (protandry). As an example, most wrasses are protogynous hermaphrodites within a haremic mating system. Hermaphroditism allows for complex mating systems. Wrasses exhibit three different mating systems: polygynous, lek-like, and promiscuous mating systems. Group spawning and pair spawning occur within mating systems. The type of spawning that occurs depends on male body size. Labroids typically exhibit broadcast spawning, releasing high amounts of planktonic eggs, which are broadcast by tidal currents; adult wrasses have no interaction with offspring. Wrasse of a particular subgroup of the Labridae family Labrini do not exhibit broadcast spawning.

Less commonly hermaphrodites can be *synchronous*, meaning they simultaneously possess both ovaries and testicles and can function as either sex at any one time. Black hamlets "take turns releasing sperm and eggs during spawning. Because such egg trading is advantageous to both individuals, hamlets are typically monogamous for short periods of time–an unusual situation in fishes." The sex of many fishes is not fixed, but can change with physical and social changes to the environment where the fish lives.

Particularly among fishes, hermaphroditism can pay off in situations where one sex is more likely to survive and reproduce, perhaps because it is larger. Anemone fishes are sequential hermaphrodites which are born as males, and become females only when they are mature. Anemone fishes live together monogamously in a anemone, protected by the anemone stings. The males do not have to compete with other males, and female anemone fish are typically larger. When a female dies a juvenile (male) anemone fish moves in, and "the resident male then turns into a female and reproductive advantages of the large female–small male combination continue". In other fishes sex changes are reversible. For example, if some gobies are grouped by sex (male or female), some will switch sex.

The mangrove rivulus *Kryptolebias marmoratus* produces both eggs and sperm by meiosis and routinely reproduces by self-fertilization. Each individual hermaphrodite normally fertilizes itself when an egg and sperm that it has produced by an internal organ unite inside the fish's body. In nature, this mode of reproduction can yield highly homozygous lines composed of individuals so genetically uniform as to be, in effect, identical to one another. The capacity for selfing in these fishes has apparently persisted for at least several hundred thousand years.

Although inbreeding, especially in the extreme form of self-fertilization, is ordinarily regarded as detrimental because it leads to expression of deleterious recessive alleles, self-fertilization does provide the benefit of *fertilization assurance* (reproductive assurance) at each generation.

Sexual Parasitism

Some anglerfish, like those of the deep sea ceratioid group, employ an unusual mating method. Because individuals are locally rare, encounters are also very rare. Therefore, finding a mate is problematic. When scientists first started capturing ceratioid anglerfish, they noticed that all of the specimens were female. These individuals were a few centimetres in size and almost all of them had what appeared to be parasites attached to them. It turned out that these "parasites" were highly reduced male ceratioids. This indicates the anglerfish use a polyandrous mating system.

The methods by which the anglerfish locate mates are variable. Some species have minute eyes unfit for identifying females visually, while others have underdeveloped nostrils, making it unlikely that they effectively find females using olfaction. When a male finds a female, he bites into her skin, and releases an enzyme that digests the skin of his mouth and her body, fusing the pair down to the blood-vessel level. The male becomes dependent on the female host for survival by receiving nutrients via their shared circulatory system, and provides sperm to the female in return. After fusing, males increase in volume and become much larger relative to free-living males of the species. They live and remain reproductively functional as long as the female lives, and can take part in multiple spawnings. This extreme sexual dimorphism ensures, when the female is ready to spawn, she has a mate immediately available. Multiple males can be incorporated into a single individual female with up to eight males in some species, though some taxa appear to have a one male per female rule.

One explanation for the evolution of sexual parasitism is that the relative low density of females in deep-sea environments leaves little opportunity for mate choice among anglerfish. Females remain large to accommodate fecundity, as is evidenced by their large ovaries and eggs. Males would be expected to shrink to reduce metabolic costs in resource-poor environments and would develop highly specialized female-finding abilities. If a male manages to find a female parasitic attachment, then it is ultimately more likely to improve lifetime fitness relative to free living, particularly when the prospect of finding future mates is poor. An additional advantage to parasitism is that the male's sperm can be used in multiple fertilizations, as he stays always available to the female for mating. Higher densities of male-female encounters might correlate with species that demonstrate facultative parasitism or simply use a more traditional temporary contact mating.

Parthenogenesis

Parthenogenesis is a form of asexual reproduction in which growth and development of embryos occur without fertilization. In animals, parthenogenesis means development of an embryo from an unfertilized egg cell. The first all-female (unisexual) reproduction in vertebrates was described in the *Amazon molly* in 1932. Since then at least 50 species of unisexual vertebrate have been described, including at least 20 fish, 25 lizards, a single snake species, frogs, and salamanders. As with all types of asexual reproduction, there are both costs (low genetic diversity and therefore susceptibility to adverse mutations that might occur) and benefits (reproduction without the need for a male) associated with parthenogenesis.

Parthenogenesis in sharks has been confirmed in the bonnethead and zebra shark. Other, usually sexual species, may occasionally reproduce parthenogenetically, and the hammerhead and blacktip sharks are recent additions to the known list of facultative parthenogenetic vertebrates.

A special case of parthenogenesis is gynogenesis. In this type of reproduction, offspring are produced by the same mechanism as in parthenogenesis, however, the egg is stimulated to develop simply by the *presence* of sperm - the sperm cells do not contribute any genetic material to the offspring. Because gynogenetic species are all female, activation of their eggs requires mating with males of a closely related species for the needed stimulus. The Amazon molly, (pictured), reproduces by gynogenesis.

Others

The elkhorn sculpin (*Alcichthys elongatus*) is a marine teleost with a unique reproductive mode called "internal gametic association". Sperm are introduced into the ovary by copulation and then enter the micropylar canal of ovulated eggs in the ovarian cavity. However, actual sperm-egg fusion does not occur until the eggs have been released into sea water.

Inbreeding

Inbreeding Depression

The effect of inbreeding on reproductive behavior was studied in the poeciliid fish *Heterandria formosa*. One generation of full-sib mating was found to decrease reproductive performance and likely reproductive success of male progeny. Other traits that displayed inbreeding depression were offspring viability and maturation time of both males and females.

Exposure of zebra fish to a chemical environmental agent, analogous to that caused by anthropogenic pollution, amplified the effects of inbreeding on key reproductive traits. Embryo viability was significantly reduced in inbred exposed fish and there was a tendency for inbred males to sire fewer offspring.

The behaviors of juvenile Coho salmon with either low or medium inbreeding were compared in paired contests. Fish with low inbreeding showed almost twice the aggressive pursuit in defending territory than fish with medium inbreeding, and furthermore had a higher specific growth rate. A significant effect of inbreeding depression on juvenile survival was also found, but only in high-density competitive environments, suggesting that intra-specific competition can magnify the deleterious effects of inbreeding.

Inbreeding Avoidance

Inbreeding ordinarily has negative fitness consequences (inbreeding depression), and as a result species have evolved mechanisms to avoid inbreeding. Numerous inbreeding avoidance mechanisms operating prior to mating have been described. However, inbreeding avoidance mechanisms that operate subsequent to copulation are less well known. In guppies, a post-copulatory mechanism of inbreeding avoidance occurs based on competition between sperm of rival males for achieving fertilisation. In competitions between sperm from an unrelated male and from a full sibling male, a significant bias in paternity towards the unrelated male was observed.

Inbreeding depression is considered to be due largely to the expression of homozygous deleterious recessive mutations. Outcrossing between unrelated individuals results in the beneficial masking of deleterious recessive mutations in progeny.

Examples

Female goldfish sp awn (discharge) eggs into the water, encouraged by male goldfish who simultaneously discharge sperm which externally fertilizes the eggs.

Within two or three days, the vulnerable goldfish eggs hatch into larvae, and rapidly develop into fry.

Goldfish

Goldfish, like all cyprinids, are egg-layers. They usually start breeding after a significant temperature change, often in spring. Males chase females, prompting them to release their eggs by bumping and nudging them. As the female goldfish spawns her eggs, the male goldfish stays close behind fertilizing them. Their eggs are adhesive and attach to aquatic vegetation. The eggs hatch within 48 to 72 hours. Within a week or so, the fry begins to assume its final shape, although a year may pass before they develop a mature goldfish colour; until then they are a metallic brown like their wild ancestors. In their first weeks of life, the fry grow quickly—an adaptation born of the high risk of getting devoured by the adult goldfish.

Carp

A member of the Cyprinidae family, carp spawn in times between April and August, largely dependent upon the climate and conditions they live in. Oxygen levels of the water, availability of food, size of each fish, age, number of times the fish has spawned before and water temperature are all factors known to effect when and how many eggs each carp will spawn at any one time.

Siamese Fighting Fish

Prior to spawning, male Siamese fighting fish build bubble nests of varying sizes at the surface of the water. When a male becomes interested in a female, he will flare his gills, twist his body, and spread his fins. The female darkens in colour and curves her body back and forth. The act of spawning takes place in a "nuptial embrace" where the male wraps his body around the female, each embrace resulting in the release of 10-40 eggs until the female is exhausted of eggs. The male, from his side, releases milt into the water and fertilization takes place externally. During and after spawning, the male uses his mouth to retrieve sinking eggs and deposit them in the bubble nest (during mating the female sometimes assists her partner, but more often she will simply devour all the eggs that she manages to catch). Once the female has released all of her eggs, she is chased away from the male's territory, as it is likely that she'll eat the eggs due to hunger. The eggs then remain in the male's care. He keeps them in the bubble nest, making sure none fall to the bottom and repairing the nest as needed. Incubation lasts for 24–36 hours, and the newly hatched larvae remain in the nest for the next 2–3 days, until their yolk sacs are fully absorbed. Afterwards the fry leave the nest and the free-swimming stage begins.

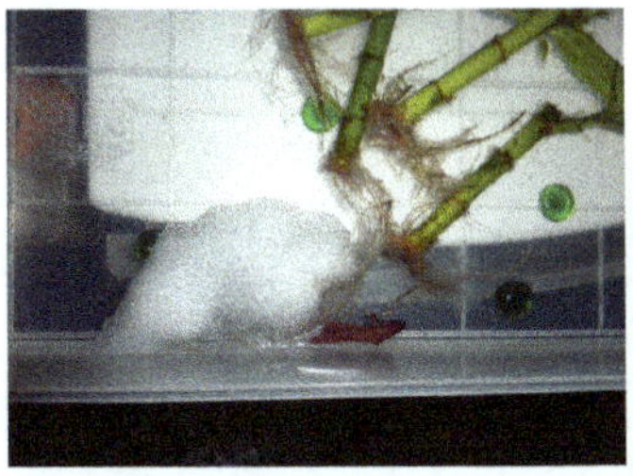		
Siamese fighting fish build bubble nests of varying sizes.	A pair of Siamese fighting fish spawning under their bubble nest.	A 15-day-old free-swimming fry of a Siamese fighting fish.

Spawn

The spawn (eggs) of a clownfish. The black spots are the developing eyes.

Spawn is the eggs and sperm released or deposited into water by aquatic animals. As a verb, *to spawn* refers to the process of releasing the eggs and sperm, and the act of both sexes is called spawning. Most aquatic animals, except for aquatic mammals and reptiles, reproduce through the process of spawning.

Spawn consists of the reproductive cells (gametes) of many aquatic animals, some of which will become fertilized and produce offspring. The process of spawning typically involves females releasing ova (unfertilized eggs) into the water, often in large quantities, while males simultaneously or sequentially release spermatozoa (milt) to fertilize the eggs.

Most fish reproduce by spawning, as do most other aquatic animals, including crustaceans such as crabs and shrimps, molluscs such as oysters and squid, echinoderms such as sea urchins and sea cucumbers, amphibians such as frogs and newts, aquatic insects such as mayflies and mosquitoes and corals, which are actually small aquatic animals—not plants. Fungi, such as mushrooms, are also said to "spawn" a white, fibrous matter that forms the matrix from which they grow.

There are many variations in the way spawning occurs, depending on sexual differences in anatomy, how the sexes relate to each other, where and how the spawn is released and whether or how the spawn is subsequently guarded.

Overview

Pacific salmon are semelparous or "big bang" spawners, which means they die shortly after spawning.

The pickled and dehydrated roe of mullet.

Marine animals, and particularly bony fish, commonly reproduce by *broadcast spawning*. This is an external method of reproduction where the female releases many unfertilised eggs into the water. At the same time, a male or many males release a lot of sperm into the water which fertilises some of these eggs. The eggs contain a drop of nutrient oil to sustain the embryo as it develops inside the egg case. The oil also provides buoyancy, so the eggs float and drift with the current. The strategy for survival of broadcast spawning is to disperse the fertilised eggs, preferably away from the coast into the relative safety of the open ocean. There the larvae develop as they consume their fat stores, and eventually hatch from the egg capsule into miniature versions of their parents. To survive, they must then become miniature predators themselves, feeding on plankton. Fish eventually encounter others of their own kind (conspecifics), where they form aggregations and learn to school.

Internally, the sexes of most marine animals can be determined by looking at the gonads. For example, male testes of spawning fish are smooth and white and account for up to 12% of the mass of the fish, while female ovaries are granular and orange or yellow, accounting for up to 70% of the fish's mass. Male lampreys, hagfish and salmon discharge their sperm into the body cavity where it is expelled through pores in the abdomen. Male sharks and rays can pass sperm along a duct into a seminal vesicle, where they store it for a while before it is expelled, while teleosts usually employ separate sperm ducts.

Externally, many marine animals, even when spawning, show little sexual dimorphism (difference in body shape or size) or little difference in colouration. Where species are dimorphic, such as sharks or guppies, the males often have penis-like intromittent organs in the form of a modified fin.

A species is semelparous if its individuals spawn only once in their lifetime, and iteroparous if its individuals spawn more than once. The term semelparity comes from the Latin *semel*, once, and *pario*, to beget, while iteroparity comes from *itero*, to repeat, and *pario*, to beget.

Semelparity is sometimes called "big bang" reproduction, since the single reproductive event of semelparous organisms is usually large and fatal to the spawners. The classic example of a semelparous animal is the Pacific salmon,which lives for many years in the ocean before swimming to the freshwater stream of its birth, spawning, and then dying. Other spawning animals which are semelparous include mayflies, squid, octopus, smelt, capelin and some amphibians. Semelparity is often associated with r-strategists. However, most fish and other spawning animals are iteroparous.

When the internal ovaries or egg masses of fish and certain marine animals are ripe for spawning they are called roe. Roe from certain species, such as shrimp, scallop, crab and sea urchins, are sought as human delicacies in many parts of the world. *Caviar* is a name for the processed, salted roe of non-fertilized sturgeon. The term *soft roe* or *white roe* denotes fish milt. Lobster roe is called *coral* because it turns bright red when cooked. Roe (reproductive organs) are usually eaten either raw or briefly cooked.

"The reproductive behaviour of fishes is remarkably diversified: they may be oviparous (lay eggs), ovoviparous (retain the eggs in the body until they hatch), or viviparous (have a direct tissue connection with the developing embryos and give birth to live young). All cartilaginous fishes—the elasmobranches (e.g., sharks, rays, and skates)—employ internal fertilization and usually lay large, heavy-shelled eggs or give birth to live young. The most characteristic features of the more primitive bony fishes is the assemblage of polyandrous (many males) breeding aggregations in open water and the absence of parental care..."

There are two main reproduction methods in fish. The first method is by laying eggs and the second by *live-bearing* (producing their young alive).

- In the first method, the female fish lays eggs either on the sea floor or on the leaves of an aquatic plant. A male fish fertilizes the eggs, and both then work together to protect the eggs/babies from danger until they can defend themselves.
- In the second method, the male fish uses its anal fin to transmit sperm into the female fish and fertilize the fish eggs. Later, the female gives live birth to her fry.

Sexual Strategies

Basic Strategies

The four basic mating systems		
	Single female	**Multiple females**
Single male	Monogamy	Polygyny
Multiple males	Polyandry	Polygynandry

Monogamy occurs when one male mates with one female exclusively. This is also called *pair spawning*. Most fish are not monogamous, and when they are, they often alternate with non-monogamous behaviours. Monogamy can occur when feeding and breeding grounds are small, when it is difficult for fish to find partners, or when both sexes look after the young. Many tropical cichlids, which rear their young together in locations where they must fiercely defend against competitors and predators are monogamous. "In some pipefishes and seahorses, development of eggs takes a long time before the female can place them in the brood pouch of a male, where they are fertilized. While the male is pregnant, the female starts a new batch of eggs, which are ready at about the same time that the male gives birth to the young from the previous mating. This close timing of development promotes monogamy, especially if the likelihood of encountering another potential mate is low."

Cutthroat trout are monogamous pair spawners.

The anglerfish *Haplophryne mollis* is polyandrous.
This female is trailing the atrophied remains of males she has encountered.

Polygyny occurs when one male gets exclusive mating rights with multiple females. In polygyny a large conspicuous male usually defends females from other males or defends a breeding site. The females choose large males that are successfully defending prime breeding sites which the females find attractive. For example, sculpin males defend "caves" underneath rocks which are suitable for the incubation of embryos.

Another way males get to mate with several females is through the use of leks. Leks are places where many fish come together, and the males display to each another. Based on these displays, each female then selects the male they want to be their mate. For example, among the cichlid *Cyrtocara eucinostomus* in Lake Malawi, up to 50,000 large and colourful males display together on a lek four kilometres long. The females, which are mouth brooders, choose which male they want to fertilize their eggs.

Polyandry occurs when one female gets exclusive mating rights with multiple males. This is not common, but it does happen among fish like clownfish that change their sex. It can also happen when males do the brooding but can cannot handle all the eggs the female produce, such as with some pipefish.

The males in some deep sea anglerfishes are much smaller than the females. When they find a female they bite into her skin, releasing an enzyme that digests the skin of their mouth and her body and fusing the pair down to the blood-vessel level. The male then slowly atrophies, losing first his digestive organs, then his brain, heart, and eyes, ending as nothing more than a pair of gonads, which release sperm in response to hormones in the female's bloodstream indicating egg release. This extreme sexual dimorphism ensures that, when the female is ready to spawn, she has a mate immediately available. A single anglerfish female can "mate" with many males in this manner.

Polygynandry occurs when multiple males mate indiscriminately with multiple females. This mutual promiscuity is the approach most commonly used by spawning animals, and is perhaps the "original fish mating system." Common examples are forage fish, such as herrings, which form huge mating shoals in shallow water. The water becomes milky with sperm and the bottom is draped with millions of fertilized eggs.

Cuckoldry

Small male bluegill sunfishes cuckold large males by adopting *sneaker* or *satellite* strategies.

Alternate male strategies which allow small males to engage in cuckoldry can develop in species where spawning is dominated by large and aggressive males. Cuckoldry is a variant of polyandry, and can occur with *sneak spawners* (sometimes called *streak spawners*). A sneak spawner is a male that rushes in to join the spawning rush of a spawning pair. A spawning rush occurs when a fish makes a burst of speed, usually on a near vertical incline, releasing gametes at the apex, followed by a rapid return to the lake or sea floor or fish aggregation. Sneaking males do not take part in courtship. In salmon and trout, for example, *jack males* are common. These are small silvery males that migrate upstream along with the standard, large, hook-nosed males and that spawn by sneaking into redds to release sperm simultaneously with a mated pair. This behaviour is an evolutionarily stable strategy for reproduction, because it is favoured by natural selection just like the "standard" strategy of large males.

Cuckoldry occurs in many fish species, including dragonets, parrotfishes and wrasses on tropical reefs and the bluegill sunfish in fresh water. Sneaker males that become too large to hide effectively become *satellite males*. With bluegill sunfish, satellite males mimic the behaviour and colouration of the females. They hover over a nest containing a pair of courting sunfish, and gradually descend to reach the pair just as they spawn. Males may need to be 6 or 7 years old to function capably as parental males, but may be able to function as sneaker or satellite males when they are as young as 2 or 3 years old. The smaller satellite and sneaker males may get mauled by the more powerful parental males, but they spawn when they are younger and they do not put energy into parental care.

Hermaphroditism

Hermaphroditism occurs when a given individual in a species possesses both male and female reproductive organs, or can alternate between possessing first one, and then the other. Hermaphroditism is common in invertebrates but rare in vertebrates. It can be contrasted with gonochorism, where each individual in a species is either male or female, and remains that way throughout their lives. Most fish are gonochorists, but hermaphroditism is known to occur in 14 families of teleost fishes.

Usually hermaphrodites are *sequential*, meaning they can switch sex, usually from female to male (protogyny). This can happen if a dominant male is removed from a group of females. The largest female in the harem can switch sex over a few days and replace the dominant male. This is found amongst coral reef fishes such as groupers, parrotfishes and wrasses. It is less common for a male to switch to a female (protandry). As an example, most wrasses are protogynous hermaphrodites within a haremic mating system. Hermaphroditism allows for complex mating systems. Wrasses exhibit three different mating systems: polygynous, lek-like, and promiscuous mating systems. Group spawning and pair spawning occur within mating systems. The type of spawning that occurs depends on male body size. Labroids typically exhibit broadcast spawning, releasing high amounts of planktonic eggs, which are broadcast by tidal currents; adult wrasses have no interaction with offspring. Wrasse of a particular subgroup of the Labridae family Labrini do not exhibit broadcast spawning.

Less commonly hermaphrodites can be *synchronous*, meaning they simultaneously possess both ovaries and testicles and can function as either sex at any one time. Black hamlets "take turns releasing sperm and eggs during spawning. Because such egg trading is advantageous to both individuals, hamlets are typically monogamous for short periods of time–an unusual situation in fishes." The sex of many fishes is not fixed, but can change with physical and social changes to the environment where the fish lives.

Particularly among fishes, hermaphroditism can pay off in situations where one sex is more likely to survive and reproduce, perhaps because it is larger. Anemone fishes are sequential hermaphrodites which are born as males, and become females only when they are mature. Anemone fishes live together monogamously in an anemone, protected by the anemone stings. The males do not have to compete with other males, and female anemone fish are typically larger. When a female dies a juvenile (male) anemone fish moves in, and "the resident male then turns into a female and reproductive advantages of the large female–small male combination continue". In other fishes sex changes are reversible. For example, if some gobies are grouped by sex (male or female), some will switch sex.

Unisexuality

Unisexuality occurs when a species is all-male or all-female. Unisexuality occurs in some fish species, and can take complex forms. *Squalius alburnoides*, a minnow found in several river basins in Portugal and Spain, appears to be an all-male species. The existence of this species illustrates the potential complexity of mating systems in fish. The species originated as a hybrid between two species, and is diploid, but not hermaphroditic. It can have triploid and tetraploid forms, including all-female forms that reproduce mainly through hybridogenesis.

It is rare to find true parthenogenesis in fishes, where females produce female offspring with no input from males. All-female species include the Texas silverside, *Menidia clarkhubbsi* as well as a complex of Mexican mollies. Parthenogenesis has been recently observed in hammerhead sharks and blacktip sharks. It is also known to occur in crayfish and amphibians.

Spawning Strategies

This section is patterned after a classification of the spawning behaviours of fish by Balon (1975, 1984) into reproductive guilds. This classification is based on how the eggs are fertilized (internal or external spawners), where the eggs are deposited (pelagic or benthic spawners), and whether and how the parents look after the eggs after spawning (bearers, guarders and nonguarders).

Nonguarders

Nonguarders do not protect their eggs and offspring after spawning.

Open Substrate Spawners

Nonguarders: Open Substrate Spawners

- Pelagic spawners
- Benthic spawners:
    - Spawners on coarse bottoms:
        - Pelagic free embryo and larvae
        - Benthic free embryo and larvae
    - Spawners on plants:
        - Obligatory
        - Nonobligatory
    - Spawners on fine substrates
- Terrestrial spawners

Pike usually spawn on vegetation flooded by high water. Their eggs adhere to the stems of plants.

Nonguarders: Brood Hiders

- Benthic spawners
- Crevice spawners
- Spawners on invertebrates
- Beach spawners

Bitterlings transfer responsibility for the care of their young to mussels.
This male bitterling is exhibiting spawning colours.

Open substrate spawners scatter their eggs in the environment. They usually spawn in shoals without complex courtship rituals, and males outnumber females.

Broadcast spawners: release their gametes (sperm and eggs) into open water for external fertilisation. There is no subsequent parental care. About 75% of coral species are broadcasters, the majority of which are hermatypic, or reef-building corals.

- Pelagic spawners: a type of broadcast spawners, spawn in the open sea, mostly near the surface. They are usually pelagic fish such as tuna and sardines. Some demersal fish leave the bottom to spawn pelagically, particularly coral reef fish such as parrotfish and wrasses. Pelagic spawning means water currents widely disperse the young. The eggs, embryos and larvae of pelagic spawners contain oil globules or have a high water content. As a result, they are buoyant and are widely dispersed by currents. The downside is that mortality is high, because they can be eaten so easily by pelagic predators or they can drift into unsuitable areas. Females compensate by spawning large numbers of eggs and extending their spawning periods. Pelagic spawners that live in or around coral reefs can spawn a small number of eggs almost daily over a period of months. These fishes have complex breeding behaviours including sex changes, harems, leks and territoriality.

- Benthic spawners: deposit their spawn on or near the bottom of the sea (or lake). They are usually demersal fish such as cod and flatfish. These species typically spawn without ceremony; they do not engage in elaborate courtship rituals. Each female is usually followed by several males who fertilize the eggs as they are released. Various strategies ensure the eggs and embryos remain in place, and do not drift with the current. The eggs can adhere to other eggs or to whatever they are deposited on, or the eggs can be laid in long strings which are wrapped around plants or rocks. Some eggs take on water after they are released, so they can be dropped into cracks where they swell and wedge themselves in place.

- Egg scatterers: scatter adhesive or non-adhesive eggs to fall to the substrate, into plants, or float to the surface. These species do not look after their brood and even eat their own eggs. These are often schooling fish which spawn in groups or pairs, often laying a large number of small eggs. The fry hatch quickly.
- Egg depositers: deposit eggs on a substrate (tank glass, wood, rocks, plants). Egg depositors usually lay fewer eggs than egg-scatterers, although the eggs are larger. Egg depositors fall into two groups: those that care for their eggs, and those that do not. Among egg depositors that care for their eggs are cichlids and some catfish. Egg depositors that care for their young can be divided into two groups: cavity spawners and open spawners.
- Cavity spawners: lay eggs in a cave or cavity. These fish form pairs and have advanced brood care where the eggs are defended and cleaned. The eggs take a few days to hatch, and the fry are often guarded by the parents. Various catfish, Cyprinidae, and killifish make up the majority. Cavity spawners can be contrasted with open (shelter) spawners, which lay their eggs on an open surface.

Brood Hiders

Brood hiders hide their eggs but do not give parental care after they have hidden them. Brood hiders are mostly benthic spawners that bury the fertilized eggs. For example, among salmon and trout the female digs a nest with her tail in gravel. These nests are called redds. The female then lays her eggs while the male fertilizes them, while both fish defend the redd if necessary from other members of the same species. Then the female buries the nest, and the nest site is abandoned. In North America, some minnows build nests out of piles of stones rather than dig holes. The minnow males have tubercles on their head and body which they use to help them defend the nest site.

- Egg buriers - can inhabit waters that dry up at some time of the year. An example are annual killifish which lay their eggs in mud. The parents mature quickly and lay their eggs before dying when the water dries up. The eggs remain in a dormant stage until rains stimulate hatching.

Bitterlings have a remarkable reproduction strategy where parents transfer responsibility for the care of their young to mussels. The female extends her ovipositor into the mantle cavity of the mussel and deposits her eggs between the gill filaments. The male then ejects his sperm into the mussel's inhalant water current and fertilization takes place within the gills of the host. The same female may use a number of mussels, and she deposits only one or two yellow, oval eggs into each. Early developmental stages are protected from predation within the body of the mussel. After 3 to 4 weeks larvae swim away from the host to continue life on their own.

Guarders

Guarders: Substrate Spawners

- Rock tenders
- Plant tenders
- Terrestrial tenders
- Pelagic tenders

Damselfish are substrate spawners. This one keeps her spawn in a gastropod shell.

Guarders: Nest Spawners

- Rock and gravel nesters
- Sand nesters
- Plant-material nesters:
    - Gluemakers
    - Nongluemakers
- Bubble nesters
- Hole nesters
- Misc-materials nesters
- Anemone nesters

Guarders protect their eggs and offspring after spawning by practicing *parental care* (also called *brood care*). Parental care is an "investment by parents in offspring that increases the offsprings' chances of surviving (and hence reproducing). In fish, parental care can take a variety of forms including guarding, nest building, fanning, splashing, removal of dead eggs, retrieval of straying fry, external egg carrying, egg burying, moving eggs or young, ectodermal feeding, oral brooding, internal gestation, brood-pouch egg carrying, etc."

The stickleback glues plant material to make its nest.

Territorial behaviour is generally necessary for guarders, and the embryos are almost always guarded by males (apart from cichlids). There is a need to be territorial because looking after embryos usually includes defending the site where they are being looked after. It also often means there is competition for the best egg-laying sites. Elaborate courtship behaviour is usual among guarders.

Guarding males keep the embryos safe from predators, keep oxygen levels high by fanning water currents, and keep the area free from dead embryos and debris. They protect the embryos until they hatch, and often look after the larval stages as well. The time spent guarding can range from a few days to several months.

Substrate Spawners

Some guarders build nests (*nest spawners*) and some do not (*substrate spawners*), though the difference between the two groups can be small. Substrate spawners clean off a suitable area of surface suitable for egg laying, and look after the area, but they do not actively build a nest.

Baby paradise fish just hatched, gathered under the surface of a bubble nest.

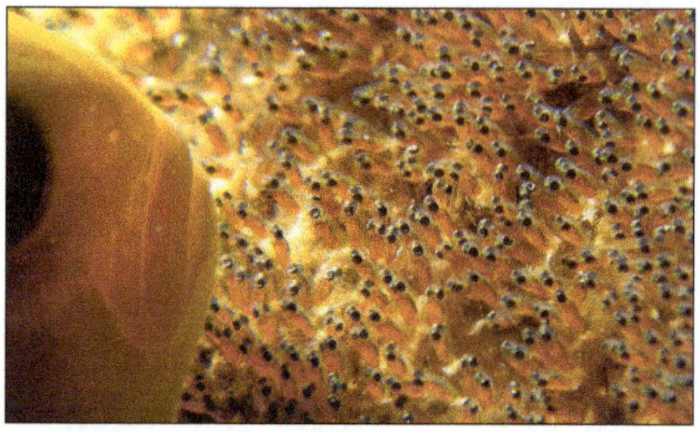

Anemone fish nest in an anemone. Here a male is protecting spawn produced by his partner.

Bearers

Bearers: External

- Transfer brooders
- Auxiliary brooders
- Mouth brooders
- Gill-chamber brooders
- Pouch brooders

Male seahorsees are pouch brooders.

A female cichlid mouthbrooding fry which can be seen looking out her mouth.

Bearers are fish that carry their embryos (and sometimes their young) around with them, either externally or internally.

External Bearers

Mouth brooders - carry eggs or larvae in their mouth. Mouth brooders can be ovophiles or larvophiles. Ovophile or egg-loving mouth-brooders lay their eggs in a pit, which are sucked up into the mouth of the female. The small number of large eggs hatch in the mother's mouth, and the fry remain there for a period of time. Fertilization often occurs with the help of egg-spots, which are colorful spots on the anal fin of the male. When the female sees these spots, she tries to pick up the egg-spots, but instead gets a mouthful of sperm, fertilizing the eggs in her mouth. Many cichlids and some labyrinth fish are ovophile mouthbrooders. Larvophile or larvae-loving mouth-brooders lay their eggs on a substrate and guard them until the eggs hatch. After hatching, the female picks up the fry and keeps them in her mouth. When the fry can fend for themselves, they are released. Some eartheaters are larvophile mouthbrooders.

Internal Bearers

Facultative Internal Bearers

The beginning of the evolutionary process of livebearing starts with *facultative* (optional) internal bearing. The process occurs in several species of oviparous (egg-laying) killifishes which spawn in the normal way on the substrate, but in the process accidentally fertilize eggs which the female retains and does not spawn. These eggs are spawned later, usually without allowing much time for embryonic development.

Bearers: Internal

- Facultative internal bearers
- Obligate internal bearers
- Livebearers

Guppies are livebearers. This one has been pregnant for about 26 days.

Obligate Internal Bearers

The next step in the evolution of livebearing is *obligate* (by necessity) internal bearing, where the female retains all the embryos. "The only source of nutrition for these embryos, however, is the egg yolk, as in externally spawned eggs. This situation, also referred to as ovoviviparity, is characteristic of marine rock fishes and the Lake Baikal sculpins. This strategy allows these fish to have fecundities approaching those of pelagic fish with external fertilization, but it also enables them to protect the young during their most vulnerable stage of development. By contrast, sharks and rays using this strategy produce a relatively small number of embryos and retain them for a few weeks to 16 months or longer. The shorter times spans are characteristic of species that eventually deposit their embryos in the environment, surrounded by a horny capsule; whereas the longer periods are characteristic of sharks that retain the embryos until they are ready to emerge as actively swimming young."

Viviparous Fish

However, some fish do not fit these categories. The livebearing largespring gambusia (*Gambusia geiseri*) was thought to be ovoviviparous until it was shown in 2001 that the embryos received nutrients from the mother.

Spawning Grounds

Capelin migrate huge distances to their spawning grounds.

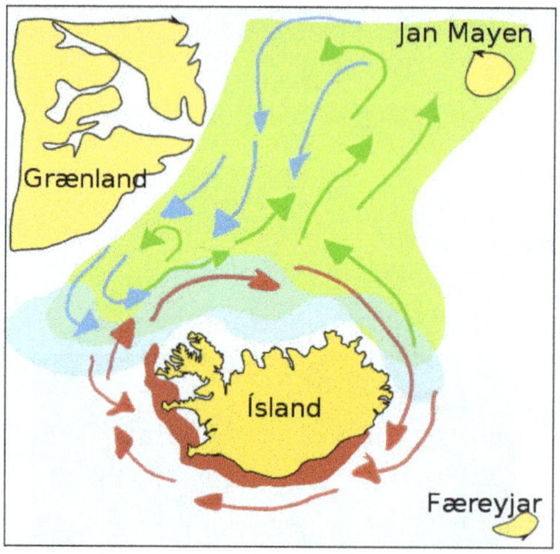

Migration of capelin around Iceland. Capelin on the way to feeding grounds is green, capelin on the way back is blue, the breeding grounds are red.

Spawning grounds are the areas of water where aquatic animals spawn, or produce their eggs. After spawning, the spawn may or may not drift to new grounds which become their nursery grounds. Many species undertake migrations each year, and sometimes great migrations, to reach their spawning grounds. For example, lakes and river watersheds can be major spawning grounds for anadromous fish such as salmon. These days, it is often necessary to construct fish ladders and other bypass systems so salmon can navigate their way past hydroelectric dams or other obstructions such as weirs on their way to spawning grounds. Coastal fish often use mangroves and estuaries as spawning grounds, while reef fish can find adjacent seagrass meadows that make good spawning grounds. Short-finned eels can travel anything up to three or four thousand kilometres to their spawning ground in deep water somewhere in the Coral Sea.

Forage fish often make great migrations between their spawning, feeding and nursery grounds. Schools of a particular stock usually travel in a triangle between these grounds. For example, one stock of herrings have their spawning ground in southern Norway, their feeding ground in Iceland, and their nursery ground in northern Norway. Wide triangular journeys such as these may be important because forage fish, when feeding, cannot distinguish their own offspring.

Capelin are a forage fish of the smelt family found in the Atlantic and Arctic oceans. In summer, they graze on dense swarms of plankton at the edge of the ice shelf. Larger capelin also eat krill and other crustaceans. The capelin move inshore in large schools to spawn and migrate in spring and summer to feed in plankton rich areas between Iceland, Greenland, and Jan Mayen. The migration is affected by ocean currents. Around Iceland maturing capelin make large northward feeding migrations in spring and summer. The return migration takes place in September to November. The spawning migration starts north of Iceland in December or January.

The diagram on the right shows the main spawning grounds and larval drift routes. Capelin on the way to feeding grounds is coloured green, capelin on the way back is blue, and the breeding grounds are red. In a paper published in 2009, researchers from Iceland recount their application of an interacting particle model to the capelin stock around Iceland, successfully predicting the spawning migration route for 2008.

Referred to as “the greatest shoal on earth”, the sardine run occurs when millions of sardines migrate from their spawning grounds south of the southern tip of Africa northward along the Eastern Cape coastline. Chinook salmon make the longest freshwater migration of any salmon, over 3,000 kilometres (1,900 mi) up the Yukon River to spawning grounds upstream of Whitehorse, Yukon. Some green sea turtles swim more than 2,600 kilometres (1,600 mi) to reach their spawning grounds.

Examples

Goldfish

Goldfish, like all cyprinids, are egg-layers. They usually start breeding after a significant temperature change, often in spring. Males chase females, prompting them to release their eggs by bumping and nudging them. As the female goldfish spawns her eggs, the male goldfish stays close behind fertilizing them. Their eggs are adhesive and attach to aquatic vegetation. The eggs hatch within 48 to 72 hours. Within a week or so, the fry begins to assume its final shape, although a year may pass before they develop a mature goldfish colour; until then they are a metallic brown like their wild ancestors. In their first weeks of life, the fry grow quickly—an adaptation born of the high risk of getting devoured by the adult goldfish.

Carp

A member of the Cyprinidae family, carp spawn in times between April and August, largely dependent upon the climate and conditions they live in. Oxygen levels of the water, availability of food, size of each fish, age, number of times the fish has spawned before and water temperature are all factors known to effect when and how many eggs each carp will spawn at any one time.

Siamese Fighting Fish

Prior to spawning, male Siamese fighting fish build bubble nests of varying sizes at the surface of the water. When a male becomes interested in a female, he will flare his gills, twist his body, and spread his fins. The female darkens in colour and curves her body back and forth. The act of spawning takes place in a "nuptial embrace" where the male wraps his body around the female, each embrace resulting in the release of 10-40 eggs until the female is exhausted of eggs. The male, from his side, releases milt into the water and fertilization takes place externally. During and after spawning, the male uses his mouth to retrieve sinking eggs and deposit them in the bubble nest (during mating the female sometimes assists her partner, but more often she will simply devour all the eggs that she manages to catch). Once the female has released all of her eggs, she is chased away from the male's territory, as it is likely that she'll eat the eggs due to hunger. The eggs then remain in the male's care. He keeps them in the bubble nest, making sure none fall to the bottom and repairing the nest as needed. Incubation lasts for 24–36 hours, and the newly hatched larvae remain in the nest for the next 2–3 days, until their yolk sacs are fully absorbed. Afterwards the fry leave the nest and the free-swimming stage begins.

One-day-old Siamese fighting fish larvae in a bubble nest - their yolk sacs have not yet been absorbed.

Crustaceans

California spiny lobster.

Copepods

Copepods are tiny crustaceans which usually reproduce either by *broadcast spawning* or by *sac spawning*. Broadcasting copepods scatter their eggs into the water, but sac spawners lay their eggs into an ovigerous sac. Sac spawners spawn few but relatively large eggs that develop slowly. By contrast, broadcast spawners spawn numerous small eggs that develop rapidly. However, the shorter hatch times that result from broadcasting are not short enough to compensate for the higher mortality compared to sac spawners. To produce a given number of hatched eggs, broadcasters must spawn more eggs than sac spawners.

Spiny Lobsters

After mating, the fertilized eggs of the California spiny lobster are carried on the female's pleopods until they hatch, with between 120,000 and 680,000 carried by a single female. The eggs begin coral red, but darken as they develop to a deep maroon. When she is carrying the eggs, the female is said to be "berried". The eggs are ready to hatch after 10 weeks, and spawning takes place from May to August. The larvae that hatch (called *phyllosoma* larvae) do not resemble the adults. Instead, they are flat, transparent animals around 14 mm (0.55 in) long, but as thin as a sheet of paper. The larvae feed on plankton, and grow through ten molts into ten further larval stages, the last of which is around 30–32 mm (1.2–1.3 in) long. The full series of larval molts takes around 7 months, and when the last stage molts, it metamorphoses into the *puerulus* state, which is a juvenile form of the adult, though still transparent. The puerulus larvae settle to the sea floor when the water is near its maximum temperature, which in Baja California is in the fall.

Egg-bearing female lobsters migrate inshore from deeper waters to hatch their eggs, though they do not have specific spawning grounds. These lobster migrations can occur in close single-file formation "lobster trains".

Molluscs

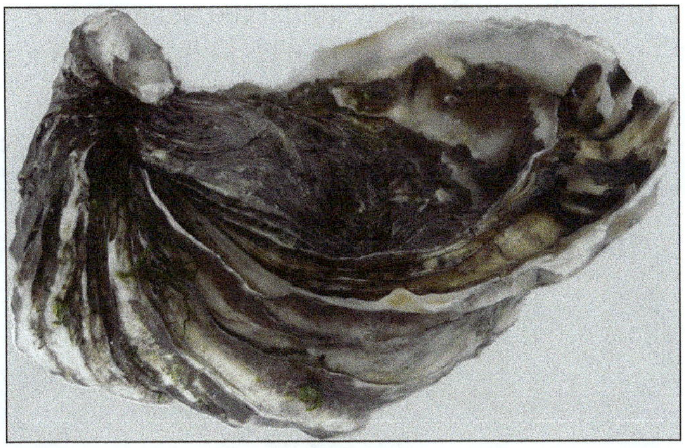

Pacific oyster

Pacific Oysters

Oysters are *broadcast spawners*, that is, eggs and sperm are released into open water where fertilisation occurs. They are protandric; during their first year they spawn as males by releasing sperm into the water. As they grow over the next two or three years and develop greater energy reserves, they spawn as females by releasing eggs. Bay oysters usually spawn by the end of June. An increase in water temperature prompts a few oysters to spawn. This triggers spawning in the rest, clouding the water with millions of eggs and sperm. A single female oyster can produce up to 100 million eggs annually. The eggs become fertilized in the water and develop rapidly into planktonic larvae. which eventually find suitable sites, such as another oyster's shell, on which to settle. Attached oyster larvae are called spat. Spat are oysters less than 25 millimetres (0.98 in) long.

The Pacific oyster usually has separate sexes. Their sex can be determined by examining the gonads, and it can change from year to year, normally during the winter months. In certain environmental conditions, one sex is favoured over the other. Protandry is favoured in areas of high food abundance and protogyny occurs in areas of low food abundance. In habitats with a high food supply, the sex ratio in the adult population tends to favour females, and areas with low food abundances tend to have a larger proportion of male adults. Spawning in the Pacific oyster occurs at 20 °C (68 °F). This species is very fecund, with females releasing about 50–200 million eggs in regular intervals (at a rate of 5–10 times a minute) in a single spawning. Once released from the gonads, the eggs move through the suprabranchial chambers (gills), are then pushed through the gill ostia into the mantle chamber, and are finally released in the water, forming a small cloud. In males, the sperm is released at the opposite end of the oyster, along with the normal exhalent stream of water. A rise in water temperature is thought to be the main cue in the initiation of spawning, as the onset of higher water temperatures in the summer results in earlier spawning in the Pacific oyster.

The larvae of the Pacific oyster are planktotrophic, and are about 70 µm at the prodissoconch 1 stage. The larvae move through the water column via the use of a larval foot to find suitable settlement locations. They can spend several weeks at this phase, which is dependent on water temperature, salinity and food supply. Over these weeks, larvae can disperse great distances by water currents before they metamorphose and settle as small spat. Similar to other oyster species, once the Pacific oyster larvae find a suitable habitat, they attach to it permanently using cement secreted from a gland in their foot. After settlement, the larvae metamorphose into juvenile spat. The growth rate is very rapid in optimum environmental conditions, and market size can be achieved in 18 to 30 months.

Egg cases laid by a female squid.

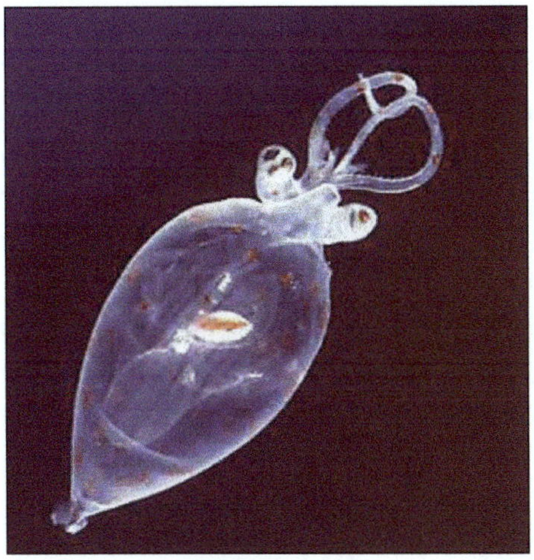

A juvenile squid.

Cephalopods

Cephalopods, such as squid and octopuses, have prominent heads and a set of arms (tentacles) modified from the primitive foot of molluscs. All cephalopods are sexually dimorphic. However, they lack external sexual characteristics, so they use colour communication. A courting male approaches a likely looking mate flashing his brightest colours, often in rippling displays. If the other cephalopod is female and receptive, her skin will change colour to become pale, and mating will occur. If the other cephalopod remains brightly coloured, it is taken as a warning.

All cephalopods reproduce by spawning eggs. Most cephalopods use semi-internal fertilization where the male places his gametes inside the female's mantle cavity to fertilize the ova in the female's single ovary. The "penis" in most male cephalopods is a long and muscular end of the gonoduct used to transfer spermatophores to a modified sperm-carrying arm called a hectocotylus. That in turn is used to transfer the spermatophores to the female. In species where the hectocotylus is missing, the "penis" is long and able to extend beyond the mantle cavity and transfers the spermatophores directly to the female. In many cephalopods, mating occurs head to head and the male may simply transfer sperm to the female. Others may detach the sperm-carrying arm and leave it attached to the female. Deep water squid have the greatest known penis length relative to body size of all mobile animals, second in the entire animal kingdom only to certain sessile barnacles. Penis elongation in the greater hooked squid may result in a penis that is as long as the mantle, head and arms combined.

A greater hooked squid with an erect penis 67 cm long.

Some species brood their fertilized eggs: female paper nautilus construct shelters for the young, while Gonatiid squid carry a larva-laden membrane from the hooks on their arms. Other cephalopods deposit their young under rocks and aerate them with their tentacles hatching. Mostly the eggs are left to their own devices; many squid lay sausage-like bunches of eggs in crevices or occasionally on the sea floor. Cuttlefish lay eggs separately in cases and attach them to coral or algal fronds. Like Pacific salmon, cephalopods are mostly semelparous, spawning many small eggs in one batch and then dying. Cephalopods usually live fast and die young. Most of the energy extracted from their food is used for growing, and they mature rapidly to their adult size. Some gain as much as 12% of their body mass each day. Most live for one to two years, reproducing and then dying shortly thereafter.

Echinoderms

Sea urchins have five gonads. These gonads (roe) are a sought after as a delicacy.

Echinoderms are marine animals, widespread in all oceans, but not found in fresh water. Just below their skin is an endoskeleton composed of calcareous plates or ossicles.

Sea Urchins

Sea urchins are spiky echinoderms with spherical bodies which usually contain five gonads. They move slowly, feed mostly on seaweed, and are important for the diet of sea otters. Sea urchins are dioecious, having separate male and female sexes, although there is generally no easy way to distinguish the two. The gonads are lined with muscles underneath the peritoneum, and these allow the animal to squeeze its gametes through the duct and into the surrounding sea water, where fertilization takes place. Their roe (male and female gonads) is soft and melting, with a colour ranging from orange to pale yellow, and is sought after as a human delicacy in many parts of the world.

Sea Cucumbers

Sea cucumbers are leathery echinoderms with elongated bodies which contain a single, branched gonad. They are found on the sea floor worldwide, and occur in great numbers on the deep sea floor where they often make up the majority of the animal biomass. They feed on plankton and decaying organic debris found at the sea bottom, catching food that flows by with their open tentacles or sifting through bottom sediments. Like sea urchins, most sea cucumbers reproduce by releasing sperm and ova into the ocean water. Depending on conditions, one organism can produce thousands of gametes.

Sea cucumbers have one gonad.

Sea cucumbers are typically dioecious, with separate male and female individuals. The reproductive system consists of a single gonad, consisting of a cluster of tubules emptying into a single duct that opens on the upper surface of the animal, close to the tentacles. Many species fertilise their eggs internally. The fertilised egg develops in a pouch on the adult's body and eventually hatches as a juvenile sea cucumber. A few species brood their young inside the body cavity, giving birth through a small rupture in the body wall close to the anus. The remaining species develop their eggs into a free-swimming larva, usually after about three days of development. This larva swims by means of a long band of cilia wrapped around its body. As the larva grows it transforms into a barrel-shaped body with three to five separate rings of cilia. The tentacles are usually the first adult features to appear, before the regular tube feet.

Amphibious Animals

Amphibians have successfully solved most of the problems associated with exposure to air. But their reproductive system was and is linked to water, and it remains very fishlike. Almost all amphibians spawn in water and lay a great number of small eggs that hatch quickly into swimming larvae. The eggs do not need any complex protection against drying, because if the environment dries, the larvae are doomed as well as the eggs. Thus selection has acted to encourage the selection of suitable sites for laying eggs, rather than suitable devices for protecting eggs. Both fishes and amphibians may migrate long distances for spawning, and favoured sites are often disputed vigorously.

Common frogs sorting out their spawn.

Frog spawn up close.

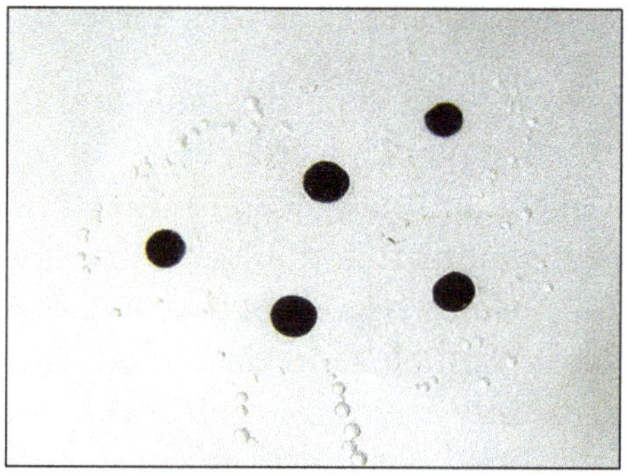

Frog spawn development.

Amphibians are found in and around fresh water lakes and ponds, but not in marine environments. Examples are frogs and toads, salamanders, newts and caecilians (which resemble snakes). They are cold-blooded animals that metamorphose from a juvenile water-breathing form, usually to an adult air-breathing form, though mudpuppies retain juvenile gills in adulthood.

Frogs and Toads

Female frogs and toads usually spawn gelatinous egg masses containing thousands of eggs in water. Different species lay eggs in distinctive and identifiable ways. For example, the American toad lays long strings of eggs. The eggs are highly vulnerable to predation, so frogs have evolved many techniques to ensure the survival of the next generation. In colder areas the embryo is black to absorb more heat from the sun, which speeds up development. Most commonly, this involves synchronous reproduction. Many individuals will breed at the same time, overwhelming the actions of predators; the majority of the offspring will still die due to predation, but there is a greater chance some will survive. Another way in which some species avoid predators and the pathogens eggs are exposed to in ponds is to lay eggs on leaves above the pond, with a gelatinous coating designed to retain moisture. In these species the tadpoles drop into the water upon hatching. The eggs of some species laid out of water can detect vibrations of nearby predatory wasps or snakes, and will hatch early to avoid being eaten.

While the length of the egg stage depends on the species and environmental conditions, aquatic eggs generally hatch within one week. Unlike salamanders and newts, frogs and toads never become sexually mature while still in their larval stage. The hatched eggs continue life as tadpoles, which typically have oval bodies and long, vertically flattened tails. As a general rule, free living larvae are fully aquatic. They lack eyelids and have a cartilaginous skeleton, a lateral line system, gills for respiration (external gills at first, internal gills later) and tails with dorsal and ventral folds of skin for swimming. They quickly develop a gill pouch that covers the gills and the front legs; the lungs are also developed at an early stage as an accessory breathing organ. Some species which go through the metamorphosis inside the egg and hatch to small frogs never develop gills; instead there are specialised areas of skin that take care of respiration. Tadpoles also lack true teeth, but the jaws in most species usually have two elongate, parallel rows of small keratinized structures called keradonts in the upper jaw while the lower jaw has three rows of keradonts, surrounded by a horny beak, but the number of rows can be lower (sometimes zero), or much higher. Tadpoles feed on algae, including diatoms filtered from the water through the gills. Some species are carnivorous at the tadpole stage, eating insects, smaller tadpoles, and fish. Cannibalism has been observed among tadpoles. Early developers who gain legs may be eaten by the others, so the late bloomers survive longer.

Sea Turtles

Sea turtle laying eggs.

Sea turtles are amphibious reptiles, but they are not amphibians. Reptiles belong to the class Reptilia while amphibians belong to the class Amphibia. These are two distinct taxonomic groups. Reptiles have scales and leathery skins, while the skins of amphibians are smooth and porous. Unlike frogs, sea turtle eggs have tough, leathery shells which allow them to survive on land without drying out.

Some sea turtles migrate long distances between feeding and spawning grounds. Green turtles have feeding grounds along the Brazilian coast. Each year, thousands of these turtles migrate about 2,300 kilometres (1,400 mi) to their spawning ground, Ascension Island in the Atlantic, an island only 11 kilometres (6.8 mi) across. Each year the returning turtles dig between 6,000 and 15,000 nests, often returning to the same beach from where they hatched. Females usually mate every two to four years. Males on the other hand visit the breeding areas every year, attempting to mate. Green sea turtles' mating is similar to other marine turtles. Female turtles control the process. A few populations practice polyandry, although this does not seem to benefit hatchlings. After mating in the water, the female moves above the beach's high tide line where she digs a hole with her hind flippers and deposits her eggs. Litter size depends on the age of the female and species, but green turtle clutches range between 100 and 200. She then covers the nest with sand and returns to the sea.

Green turtle hatchling.

At around 45 to 75 days, the eggs hatch during the night and the hatchlings instinctively head directly into the water. This is the most dangerous time in a turtle's life. As they walk, predators such as gulls and crabs grab them. A significant percentage never make it to the ocean. Little is known of the initial life history of newly hatched sea turtles. Juveniles spend three to five years in the open ocean before they settle as still-immature juveniles into their permanent shallow-water lifestyle. It is speculated that they take twenty to fifty years to reach sexual maturity. Individuals live up to eighty years in the wild. They are among the larger sea turtles, many more than a meter long and weighing up to 300 kilograms (660 lb).

Aquatic Insects

Aquatic insects also spawn. Mayflies "are famed for their short adult life. Some species have under an hour to mate and lay their eggs before they die. Their pre-adult stage, known as the subimago, may be even shorter - perhaps lasting just a few minutes before they moult into their adult form. Therefore a mayfly spends most of its life as a nymph, hidden from view under the water."

Corals

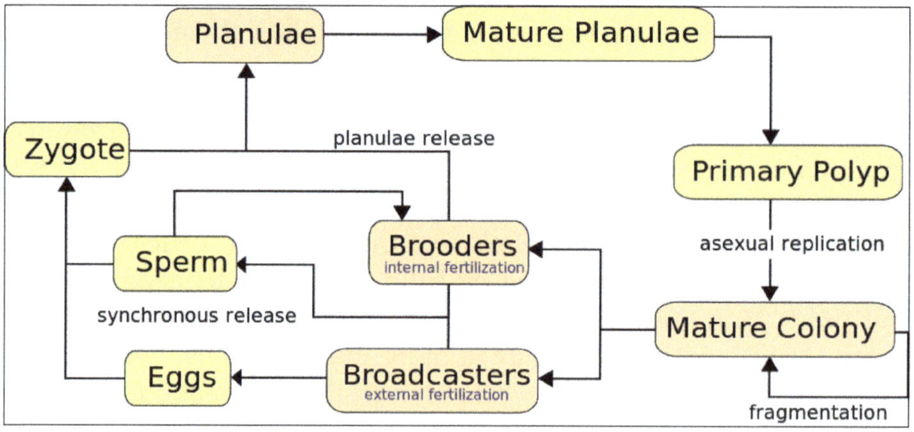

Life cycles of broadcasters and brooders.

Corals can be both gonochoristic (unisexual) and hermaphroditic, each of which can reproduce sexually and asexually. Reproduction also allows corals to settle new areas.

Corals predominantly reproduce sexually. 25% of hermatypic corals (stony corals) form single sex (gonochoristic) colonies, while the rest are hermaphroditic. About 75% of all hermatypic corals "broadcast spawn" by releasing gametes—eggs and sperm—into the water to spread offspring. The gametes fuse during fertilization to form a microscopic larva called a planula, typically pink and elliptical in shape. A typical coral colony form several thousand larvae per year to overcome the odds against formation of a new colony.

Planulae exhibit positive *phototaxis*, swimming towards light to reach surface waters where they drift and grow before descending to seek a hard surface to which they can attach and establish a new colony. They also exhibit positive *sonotaxis*, moving towards sounds that emanate from the reef and away from open water. High failure rates afflict many stages of this process, and even though millions of gametes are released by each colony very few new colonies form. The time from spawning to settling is usually 2–3 days, but can be up to 2 months. The larva grows into a polyp and eventually becomes a coral head by asexual budding and growth.

A male star coral, *Montastraea cavernosa*, releases sperm into the water.

Synchronous spawning is very typical on the coral reef and often, even when multiple species are present, all corals spawn on the same night. This synchrony is essential so that male and female gametes can meet. Corals must rely on environmental cues, varying from species to species, to determine the proper time to release gametes into the water. The cues involve lunar changes, sunset time, and possibly chemical signalling. Synchronous spawning may form hybrids and is perhaps involved in coral speciation. In some places the spawn can be visually dramatic, clouding the usually clear water with gametes, typically at night.

Corals use two methods for sexual reproduction, which differ in whether the female gametes are released:

- Broadcasters, the majority of which mass spawn, rely heavily on environmental cues, because they release both sperm and eggs into the water. The corals use long-term cues such as day length, water temperature, and/or rate of temperature change. The short-term cue is most often the lunar cycle, with sunset cuing the release. About 75% of coral species are broadcasters, the majority of which are hermatypic, or reef-building corals. The positively buoyant gametes float towards the surface where fertilization produces planula larvae. The larvae swim towards the surface light to enter into currents, where they usually remain for two days, but sometimes up to three weeks, and in one known case two months, after which they settle and metamorphose into polyps and form colonies.

- Brooders are most often ahermatypic (non-reef building) in areas of high current or wave action. Brooders release only sperm, which is negatively buoyant, and can harbor unfertilized eggs for weeks, lowering the need for mass synchronous spawning events, which do sometimes occur. After fertilization the corals release planula larvae which are ready to settle.

Fungi

Harvesting oyster mushroom *Pleurotus ostreatus* cultivated using spawns embedded in sawdust mixture placed in plastic containers.

Fungi are not plants, and require different conditions for optimal growth. Plants develop through photosynthesis, a process that converts atmospheric carbon dioxide into carbohydrates, especially cellulose. While sunlight provides an energy source for plants, mushrooms derive all of their energy and growth materials from their growth medium, through biochemical decomposition processes. This does not mean that light is an unnecessary requirement, since some fungi use light as a signal to induce fruiting. However, all the materials for growth must already be present in the growth medium. Instead of seeds, mushrooms reproduce sexually during underground growth, and asexually through spores. Either of these can be contaminated with airborne microorganisms, which will interfere with mushroom growth and prevent a healthy crop. Mycelium, or actively growing mushroom culture, is placed on growth substrate to seed or introduce mushrooms to grow on a substrate. This is also known as inoculation, spawning or adding spawn. Its main advantages are to reduce chances of contamination while giving mushrooms a firm beginning.

Ichthyoplankton

Ichthyoplankton (from Greek, *ikhthus*, "fish" and *planktos*, "drifter") are the eggs and larvae of fish. They are usually found in the sunlit zone of the water column, less than 200 metres deep, which is sometimes called the epipelagic or photic zone. Ichthyoplankton are planktonic, meaning they cannot swim effectively under their own power, but must drift with the ocean currents. Fish eggs cannot swim at all, and are unambiguously planktonic. Early stage larvae swim poorly, but later stage larvae swim better and cease to be planktonic as they grow into juveniles. Fish larvae are part of the zooplankton that eat smaller plankton, while fish eggs carry their own food supply. Both eggs and larvae are themselves eaten by larger animals.

Fish can produce high numbers of eggs which are often released into the open water column. Fish eggs typically have a diameter of about 1 millimetre (0.039 in). The newly hatched young of oviparous fish are called larvae. They are usually poorly formed, carry a large yolk sac (for nourishment) and are very different in appearance from juvenile and adult specimens. The larval period in oviparous fish is relatively short (usually only several weeks), and larvae rapidly grow and change appearance and structure (a process termed metamorphosis) to become juveniles. During this transition larvae must switch from their yolk sac to feeding on zooplankton prey, a process which depends on typically inadequate zooplankton density, starving many larvae.

Ichthyoplankton can be a useful indicator of the state and health of an aquatic ecosystem. For instance, most late stage larvae in ichthyoplankton have usually been preyed on, so ichthyoplankton tends to be dominated by eggs and early stage larvae. This means that when fish, such as anchovies and sardines, are spawning, ichthyoplankton samples can reflect their spawning output and provide an index of relative population size for the fish. Increases or decreases in the number of adult fish stocks can be detected more rapidly and sensitively by monitoring the ichthyoplankton associated with them, compared to monitoring the adults themselves. It is also usually easier and more cost effective to sample trends in egg and larva populations than to sample trends in adult fish populations.

History

Interest in plankton originated in Britain and Germany in the nineteenth century when researchers discovered there were microorganisms in the sea, and that they could trap them with fine-mesh nets. They started describing these microorganisms and testing different net configurations. Ichthyoplankton research started in 1864 when the Norwegian government commissioned the marine biologist G. O. Sars to investigate fisheries around the Norwegian coast. Sars found fish eggs, particularly cod eggs, drifting in the water. This established that fish eggs could be pelagic, living in the open water column like other plankton. Around the beginning of the twentieth century, research interest in ichthyoplankton became more general when it emerged that, if ichthyoplankton was sampled quantitatively, then the samples could indicate the relative size or abundance of spawning fish stocks.

Sampling Methods

Retrieving a Plankton Sample

Research vessels collect ichthyoplankton from the ocean using fine mesh nets. The vessels either tow the nets through the sea or pump sea water onboard and then pass it through the net.

- There are many types of plankton tows:
  - Neuston net tows are often made at or just below the surface using a nylon mesh net fitted to a rectangular frame.

- The PairoVET tow, used for collecting fish eggs, drops a net about 70 metres into the sea from a stationary research vessel and then drags it back to the vessel.
- Ring net tows involve a nylon mesh net fitted to a circular frame. These have largely been replaced by bongo nets, which provide duplicate samples with their dual-net design.
- The bongo tow drags nets shaped like bongo drums from a moving vessel. The net is often lowered to about 200 metres and then allowed to rise to the surface as it is towed. In this way, a sample can be collected across the whole photic zone where most ichthyoplankton is found.
- MOCNESS tows and Tucker trawls utilize multiple nets that are mechanically opened and closed at discrete depths in order to provide insights into the vertical distribution of the plankton.
- The manta trawl tows a net from a moving vessel along the surface of the water, collecting larvae, such as grunion, mahi-mahi, and flying fish which live at the surface.

After the tow the plankton is flushed with a hose to the cod end (bottom) of the net for collection. The sample is then placed in preservative fluid prior to being sorted and identified in a laboratory.

- Plankton pumps: Another method of collecting ichthyoplankton is to use a Continuous Underway Fish Egg Sampler (CUFES). Water from a depth of about three metres is pumped onto the vessel and filtered with a net. This method can be used while the vessel is underway.

Developmental Stages

Ichthyoplankton researchers generally use the terminology and development stages introduced in 1984 by Kendall and others. This consists of three main developmental stages and two transitional stages.

Survival

Recruitment of fish is regulated by larval fish survival. Survival is regulated by prey abundance, predation, and hydrology. Fish eggs and larvae are eaten by many marine organisms. For example, they may be fed upon by marine invertebrates, such as copepods, arrow worms, jellyfish, amphipods, marine snails and krill. Because they are so abundant, marine invertebrates inflict high overall mortality rates. Adult fish also prey on fish eggs and larvae. For example, haddock were observed satiating themselves with herring eggs back in 1922. Another study found cod in a herring spawning area with 20,000 herring eggs in their stomachs, and concluded that they could prey on half of the total egg production. Fish also cannibalise their own eggs. For example, separate studies found northern anchovy (*Engraulis mordax*) were responsible for 28% of the mortality in their own egg population, while Peruvian anchoveta were responsible for 10% and South African anchovy (*Engraulis encrasicolus*) 70%.

The most effective predators are about ten times as long as the larvae they prey on. This is true regardless of whether the predator is a crustacean, a jellyfish, or a fish.

Dispersal

The larvae of the yellow tang can drift more than 100 miles and reseed in a distant location.

Fish larvae develop first an ability to swim up and down the water column for short distances. Later they develop an ability to swim horizontally for much longer distances. These swimming developments affect their dispersal.

In 2010, a group of scientists reported that fish larvae can drift on ocean currents and reseed fish stocks at a distant location. This finding demonstrates, for the first time, what scientists have long suspected but have never proven, that fish populations can be connected to distant populations through the process of larval drift.

The fish they chose to investigate was the yellow tang, because when a larva of this fish find a suitable reef it stays in the general area for the rest of its life. Thus, it is only as drifting larvae that the fish can migrate significant distances from where they are born. The tropical yellow tang is much sought after by the aquarium trade. By the late 1990s, their stocks were collapsing, so in an attempt to save them nine marine protected areas (MPAs) were established off the coast of Hawaii. Now, through the process of larval drift, fish from the MPAs are establishing themselves in different locations, and the fishery is recovering. "We've clearly shown that fish larvae that were spawned inside marine reserves can drift with currents and replenish fished areas long distances away," said one of the authors, the marine biologist Mark Hixon. "This is a direct observation, not just a model, that successful marine reserves can sustain fisheries beyond their borders."

Pregnancy in Fish

Pregnancy has been traditionally defined as the period during which developing embryos are incubated in the body after egg-sperm union. Although the term often refers to placental mammals, it has also been used in the titles of many international, peer-reviewed, scientific articles on fish, e.g. Consistent with this definition, there are several modes of reproduction in fish, providing different amounts of parental care. In ovoviviparity, there is internal fertilization and the young are born live but there is no placental connection or significant trophic (feeding) interaction; the mother's body maintains gas exchange but the unborn young are nourished by egg yolk. There are two types of viviparity in fish. In histotrophic viviparity, the zygotes develop in the female's oviducts, but she provides no direct nutrition; the embryos survive by eating her eggs or their unborn siblings. In hemotrophic viviparity, the zygotes are retained within the female and are provided with nutrients by her, often through some form of placenta.

A pregnant fish.

In seahorses and pipefish, it is the male that becomes pregnant.

Types of Reproduction and Pregnancy

Pregnancy has been traditionally defined as the period during which developing embryos are incubated in the body after egg-sperm union. Despite strong similarities between viviparity in mammals, researchers have historically been reluctant to use the term "pregnancy" for non-mammals because of the highly developed form of viviparity in eutherians. Recent research into physiological, morphological and genetic changes associated with fish reproduction provide evidence that incubation in some species is a highly specialized form of reproduction similar to other forms of viviparity. Although the term "pregnancy" often refers to eutherian animals, it has also been used in the titles of many international, peer-reviewed, scientific articles on fish, e.g. five modes of reproduction can be differentiated in fish based on relations between the zygote(s) and parents:

- Ovuliparity: Fertilization of eggs is external; zygotes develop externally.
- Oviparity: Fertilization of eggs is internal; zygotes develop externally as eggs with large vitellus.
- Ovoviviparity: Fertilization is internal; zygotes are retained in the female (or male) but without major trophic (feeding) interactions between zygote and parents (there may be minor interactions, such as maintenance of water and oxygen levels). The embryos depend upon their yolk for survival.

There are two types of viviparity among fish.

- Histotrophic ("tissue eating") viviparity: The zygotes develop in the female's oviducts, but she provides no direct nutrition. The embryos survive by eating her eggs or their unborn siblings.
- Hemotrophic ("blood eating") viviparity: The zygotes are retained within the female and are provided with nutrients by her, often through some form of placenta.

Ovoviviparous Fish

Examples of ovoviviparous fish are many of the squaliform sharks, which include sand sharks, mackerel sharks, nurse sharks, requiem sharks, dog sharks and hammerheads, among others, and the lobe finned coelacanth. Some species of rockfish (*Sebastes*) and sculpins (Comephoridae) produce rather weak larvae with no egg membrane and are also, by definition, ovoviviparous. Ovoviparity occurs in most live-bearing bony fishes (Poeciliidae).

Viviparous Fish

Viviparous fish include the families Goodeidae, Anablepidae, Jenynsiidae, Poeciliidae, Embiotocidae and some sharks (some species of the requiem sharks, Carcharinidae and the hammerheads, Sphyrnidae, among others). The halfbeaks, Hemiramphidae, are found in both marine and freshwaters and those species that are marine produce eggs with extended filaments that attach to floating or stationary debris, while those that are found in freshwater are viviparous with internal fertilization. The Bythitidae are also viviparous although one species, *Dinematichthys ilucoeteoides*, is reported to be ovoviviparous.

Aquarists commonly refer to ovoviviparous and viviparous fish as "livebearers". Examples include guppies, mollies, moonfish, platys, four-eyed fish and swordtails. All of these varieties exhibit signs of their pregnancy before the live fry are born. As examples, the female swordtail and guppy will both give birth to anywhere from 20 to 100 live young after a gestation period of four to six weeks, and mollies will produce a brood of 20 to 60 live young after a gestation of six to 10 weeks.

Nutrition during Pregnancy

Other terms relating to pregnancy in fish relate to the differences in the mode and extent of support the female gives the developing offspring.

"Lecithotrophy" (yolk feeding) occurs when the mother provisions the oocyte with all the resources it needs prior to fertilization, so the egg is independent of the mother. Many members of the fish family Poeciliidae are considered to be lecithotrophic, however, research is increasingly showing that others are matrotrophic.

"Aplacentral viviparity" occurs when the female retains the embryos during the entire time of development but without any transfer of nutrients to the young. The yolk sac is the only source of nutrients for the developing embryo. There are at least two exceptions to this; some sharks gain nourishment by eating unfertilized eggs produced by the mother (oophagy or egg eating) or by eating their unborn siblings (intra-uterine cannibalism).

"Matrotrophy" (mother feeding) occurs when the embryo exhausts its yolk supply early in gestation and the mother provides additional nutrition. Post-fertilization transfer of nutrients has been reported in several species within the genera *Gambusia* and *Poecilia*, specifically, *G. affinis*, *G. clarkhubbsi*, *G. holbrooki*, *G. gaigei*, *G. geiseri*, *G. nobilis*, *P. formosa*, *P. latipinna*, and *P. mexicana*.

Viviparous fish have developed several ways of providing their offspring with nutrition. "Embryotrophic" or "histrotrophic" nutrition occurs by the production of nutritive fluid, uterine milk, by the uterine lining, which is absorbed directly by the developing embryo. "Hemotrophic" nutrition occurs through the passing of nutritive substances between blood vessels of the mother and embryo that are in close proximity, i.e. a placenta-like organ similar to that found in mammals.

Comparison between Species

There is considerable variation between species in the length of pregnancy. At least one group of fish has been named after its pregnancy characteristics. The surfperch, genus *Embiotoca*, is a saltwater fish with a gestation period of three to six months. This lengthy period of pregnancy gives the family its scientific name from the Greek "embios" meaning "persistent" and "tokos" meaning "birth".

The table below shows the gestation period and number of young born for some selected fish.

Species		Reproduction method	Gestation period (Days)	Number of young (Average)
Atlantic sharpnose shark	(*Rhizoprionodon terraenovae*)	Viviparous	300-330	4-6
Barbeled houndshark	(*Leptocharias smithii*)	Viviparous[a]	>120	7
Blackspot shark	(*Carcharhinus sealei*)	Viviparous[b]	270	1-2
Blue shark	(*Prionace glauca*)	Viviparous	270-366	4-135
Bonnethead shark	(*Sphyrna tiburo*)	Viviparous[c]		4-12
Bull shark	(*Carcharhinus leucas*)	Viviparous	366	4-10
Butterfly goodeid	(*Ameca splendens*)	Viviparous	55-60	6-30
Caribbean sharpnose shark	(*Rhizoprionodon porosus*)	Viviparous		2-6
Daggernose shark	(*Isogomphodon oxyrhynchus*)	Viviparous	366	2-8
Lemon shark	(*Negaprion brevirostris*)	Viviparous	366	18 (max)
Oceanic whitetip shark	(*Carcharhinus longimanus*)	Viviparous	366	1-15
Dwarf seahorse	(*Hippocampus zosterae*)	Viviparous	3-55	10
Sandbar shark	(*Carcharhinus plumbeus*)	Viviparous	366	8
Spadenose shark	(*Scoliodon laticaudus*)	Viviparous[d]	150-180	6-18
Viviparous eelpout	(*Zoarces viviparus*)	Viviparous[e]	180	30-400
Basking shark	(*Cetorhinus maximus*)	Ovoviviparous	>366	unknown[f]
Bat ray	(*Myliobatis californica*)	Ovoviviparous	270-366	2-10
Coelacanth	(g. *Latimeria*)	Ovoviviparous	>366	
Blue stingray	(*Dasyatis chrysonota*)	Ovoviviparous	270	1-5
Bluespotted stingray	(*Neotrygon kuhlii*)	Ovoviviparous	90-150	1-7
Carpet sharks	(f. Ginglymostomatidae)	Ovoviviparous	180	30-40
Knifetooth sawfish	(*Anoxypristis cuspidata*)	Ovoviviparous	150	6-23
Nurse shark	(*Ginglymostoma cirratum*),	Ovoviviparous	150	21-29
Sailfin molly	(*Poecilia latipinna*)	Ovoviviparous	21-28	10-140
Salmon shark	(*Lamna ditropis*)	Ovoviviparous	270	2-6
Sand tiger shark	(*Carcharias taurus*)	Ovoviviparous	270-366	2[g]
School shark	(*Galeorhinus galeus*)	Ovoviviparous	366	28-38
Shortfin mako shark	(*Isurus oxyrinchus*)	Ovoviviparous	450-540	4-18
Spotted eagle ray	(*Aetobatus narinari*)	Ovoviviparous	366	4
Tiger shark	(*Galeocerdo cuvier*)	Ovoviviparous	480	10-80
Tawny nurse shark	(*Nebrius ferrugineus*)	Aplacental viviparity		1-2

- Unlike any other shark, the yolk-sac placenta is globular or spherical.

- At first, the embryos are sustained by a yolk sac, but later a placenta develops.
- A bonnethead female produced a pup by parthenogenesis in 2001.
- The spadenose shark has the most advanced form of placental viviparity known in fish, as measured by the complexity of the placental connection and the difference in weight between the egg and the newborn young.
- The eelpout suckles its young embryos while still within their mother's body, making it the only fish species to suckle its offspring.
- Only one pregnant female is known to have been caught; she was carrying six unborn young.
- 1 per uterine horn.

Poeciliopsis

Members of the genus *Poeciliopsis* (amongst others) show variable reproductive life history adaptations. *P. monacha* can be considered to be lecithotrophic because the female does not really provide any resources for her offspring after fertilization. *P. lucida* shows an intermediate level of matrotrophy, meaning that to a certain extent, the offspring's metabolism can actually affect the mother's metabolism, allowing for increased nutrient exchange. *P. prolifica* is considered to be highly matrotrophic, and almost all of the nutrients and materials needed for foetal development are supplied to the oocyte after it has been fertilized. This level of matrotrophy allows *Poeciliopsis* to carry several broods at different stages of development, a phenomenon known as superfetation.

P. elongata, *P. turneri* and *P. presidionis* form another clade which could be considered an outgroup to the *P. monacha*, *P.lucida*, and *P. prolifica* clade. These three species are very highly matrotrophic – so much so that in 1947, C. L. Turner described the follicular cells of *P. turneri* as "pseudo-placenta, pseudo-chorion, and pseudo-allantois".

Guppy

Guppies are highly prolific livebearers giving birth to between five and 30 fry, though under extreme circumstances, she may give birth to only one or two or over 100. The gestation period of a guppy is typically 21–30 days, but can vary considerably. The area where a pregnant guppy's abdomen meets the tail is sometimes called the "gravid patch", or "gravid spot". When pregnant, there is a slight discoloration that slowly darkens as the guppy progresses through pregnancy. The patch first has a yellowish tinge, then brown and become deep orange as the pregnancy develops. This patch is where the fertilized eggs are stored and grow. The darkening is actually the eyes of the developing baby guppies and the orange tinge is their jelly-like eggs.

Elasmobranchs

The majority of elasmobranchs are viviparous and show a wide range of strategies to provide their offspring with nourishment and respiratory requirements. Some sharks simply retain their young in the dilated posterior segment of the oviduct. In its simplest form, the uterus does not provide any additional nutrients to the embryos. However, other elasmobranchs develop secretory uterine villi that produce histotroph, a nutrient which supplements the yolk stores of the oocyte. Uterine secretions are perhaps most advanced in the stingrays. Following depletion of the yolk, the uterine lining hypertrophies into secretory appendages termed "trophonemata". The process by which the uterine secretions (also known as uterine milk or histotroph) are produced resembles that of breast milk in mammals. Furthermore, the milk is rich in protein and lipid. As the embryo grows, vascularisation of the trophonemata enlarges to form sinusoids that project out to the surface to form a functional respiratory membrane. In lamnoid sharks, following yolk use, the embryos develop teeth and eat eggs and siblings within the uterus. There is usually one fetus per uterus and it grows to enormous proportions of up to 1.3 m in length. In placental sharks, the yolk sac is not withdrawn to become incorporated into the abdominal wall. Rather, it lengthens to form an umbilical cord and the yolk sac becomes modified into a functional epitheliochorial placenta.

Male Pregnancy

Pregnant male seahorse

The male fishes of seahorses, pipefishes, weedy and leafy sea dragons (Syngnathidae) are unusual as the male, rather than the female, incubates the eggs before releasing live fry into the surrounding water. To achieve this, male seahorses protect eggs in a specialized brood pouch, male sea dragons attach their eggs to their legs, and male pipefish may do either.

When a female's eggs reach maturity, she squirts them from a chamber in her trunk via her ovipositor into his brood pouch or egg pouch, sometimes called a "marsupium". During a mammalian pregnancy, the placenta allows the female to nourish her progeny in the womb, and remove their waste products. If male pipefish and seahorses provide only a simple pouch for fish eggs to develop and hatch, it might not fully qualify as bona-fide pregnancy. However, current research suggests that in syngnathid species with well developed brood pouches, males do provide nutrients, osmoregulation and oxygenation to the embryos they carry.

Seahorse

When mating, the female seahorse deposits up to 1,500 (average of 100 to 1,000) eggs in the male's pouch, located on the ventral abdomen at the base of the tail. Male juveniles develop pouches when they are 5–7 months old. The male carries the eggs for 9 to 45 days until the seahorses emerge fully developed, but very small. The number born maybe as few as five for smaller species, or 2,500 for larger species. A male seahorse's body has large amounts of prolactin, the same hormone that governs milk production in pregnant mammals and although the male seahorse does not supply milk, his pouch provides oxygen as well as a controlled-environment.

When the fry are ready to be born, the male expels them with muscular contractions, sometimes while attaching himself to seaweed with his tail. Birth typically occurs during the night, and a female returning for the routine morning greeting finds her mate ready for the next batch of eggs.

Pipefish

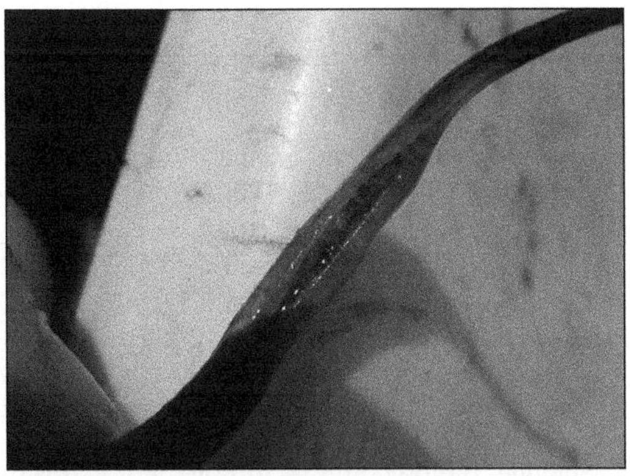

The subcaudal pouch of the male black-striped pipefish.

Pipefish brood their offspring either on distinct region of its body or in a brood pouch. Brood pouches vary significantly among different species of pipefish, but all contain a small opening through which the female's eggs can be deposited. The location of the brood pouch can be along the entire underside of the pipefish or just at the base of the tail, as with seahorses. Pipefish in the genus *Syngnathus* have a brood pouch with a ventral seam that can completely cover all of their eggs when sealed. In males without these pouches, eggs adhere to a strip of soft skin on the ventral surface of their bodies that does not contain any exterior covering – a type of "skin brooding".

At least two species of pipefish, *Syngnathus fuscus* and *Syngnathus floridae*, provide nutrients for their offspring.

The table below shows the gestation period and number of young born for some selected seahorses.

Species		Reproduction method	Gestation period (Days)	Number of young
Big-belly seahorse	(*Hippocampus abdominalis*)	Ovoviviparous	28	600-700
Lined seahorse	(*Hippocampus erectus*)	Ovoviviparous	20-21	650 (max)
Long-snouted seahorse	(*Hippocampus guttulatus*)	Ovoviviparous	21	581 (max)

References

- Stölting, K.N. & Wilson, A.B. (2007). "Male pregnancy in seahorses and pipefish: beyond the mammalian model". BioEssays. 29 (9): 884–896. PMID 17691105. doi:10.1002/bies.20626
- Chapman, D. D.; Firchau, B.; Shivji, M. S. (2008). "Parthenogenesis in a large-bodied requiem shark, the blacktip". Journal of Fish Biology. 73 (6): 1473–1477. doi:10.1111/j.1095-8649.2008.02018.x
- Lodé, T. (2012). "Oviparity or viviparity? That is the question...". Reproductive Biology. 12 (3): 259–264. doi:10.1016/j.repbio.2012.09.001. Retrieved November 4, 2014
- Compagno, L.J.V. (2002). Sharks of the World: An Annotated and Illustrated Catalogue of Shark Species Known to Date (Volume 2). Rome: Food and Agriculture Organization. ISBN 92-5-104543-7
- Helfman G, Collette BB, Facey DH and Bowen BW (2009) The Diversity of Fishes: Biology, Evolution, and Ecology p. 35, Wiley-Blackwell. ISBN 978-1-4051-2494-2
- Pierce, S. J.; Pardo, S. A.; Bennett, M. B. (2009). "Reproduction of the blue-spotted maskray Neotrygon kuhlii (Myliobatoidei: Dasyatidae) in south-east Queensland, Australia". Journal of Fish Biology. 74 (6): 1291–308. PMID 20735632. doi:10.1111/j.1095-8649.2009.02202.x
- Theodore W. Pietsch. "Precocious sexual parasitism in the deep sea ceratioid anglerfish, Cryptopsaras couesi Gill". Nature. 256: 38–40. doi:10.1038/256038a0. Archived from the original on 28 August 2008. Retrieved 31 July 2008
- Robertson, D.R.; R.R. Warner. "Sexual patterns in the labroid fishes of the Western Caribbean II: the parrotfishes (Scaridae)". Smithsonian Contributions to Zoology. 255: 1–26. doi:10.5479/si.00810282.255
- Blaber, Stephen J. M. (2000) Tropical estuarine fishes: ecology, exploitation and conservation John Wiley and Sons, Page 153–156. ISBN 978-0-632-05655-2
- Seebacher, F.; Ward, A.J.W & Wilson, R.S. (2013). "Increased aggression during pregnancy comes at a higher metabolic cost". Journal Experimental Biology. 216: 771–776. doi:10.1242/jeb.079756. Retrieved November 2, 2014
- Moe, M. "The breeder's net: Science, biology, and terminology of fish reproduction: Reproductive modes and strategies-Part 1". Advanced Aquarist. Retrieved November 1, 2014
- Jones, A.G. & Avise, J.C. (2003). "Male pregnancy". Current Biology. 13 (20): R791. PMID 14561416. doi:10.1016/j.cub.2003.09.045. Retrieved November 1, 2014
- Carrier, J.C.; Musick, J.A.; Heithaus, M.R., eds. (2012). Biology of Sharks and Their Relatives. CRC Press. pp. 296–301. ISBN 1439839247
- Hanel, R.; M. W. Westneat; C. Sturmbauer (December 2002). "Phylogenetic relationships, evolution of brood-care behavior, and geographic speciation in the Wrasse tribe Labrini". Journal of Molecular Evolution. 55 (6): 776–789. PMID 12486536. doi:10.1007/s00239-002-2373-6

Chapter 5

Fish Diseases and Parasites

Fishes can suffer from numerous diseases and parasites. Some of the common diseases of fishes can be categorized as bacterial, fungal and viral diseases. This chapter closely examines these types of fish diseases as well as the parasites which can cause diseases in fishes.

Fish Disease

Fish, like any animal, are exposed and susceptible to a wide range of diseases and parasites. In fact, it is unusual to find a fish completely free of disease organisms. It is normal to see a few dead fish from time to time. These fish usually die as the result of natural causes. Severe losses of fish due to disease organisms are rare in a natural environment. Most severe losses occur when fish are stressed due to unfavorable conditions, such as poor water quality.

Disease problems are most noticeable in the spring when fish are in a weakened state after winter and during the stressful spawning period. Also, as the water warms in spring, living conditions are better for growth of disease organisms.

There are no practical, and few safe methods, for treating disease problems in ponds. Disease outbreaks in ponds simply must run their course. Prevention or repression of disease is best accomplished by stocking only healthy fish from reputable sources and by maintaining a favorable pond environment. With the exception of the broad fish tapeworm, fish parasites are not harmful to humans. Proper cooking destroys parasites, including tapeworms.

Bacterial Diseases of Fish

Bacterial diseases are common in all fish and occur most often when environmental conditions, such as water quality, are not favorable. Inadequate oxygen levels in the pond can stress fish and make them susceptible to bacterial infections. These infections are often associated with spring die-offs in fish ponds. As the water warms during the spring, fish weakened by the winter months are often invaded by harmful bacterial that can cause death. This weakened condition can also be enhanced by frenzied spawning activity that further stresses the fish.

Bacterial losses are one of the most commonly noted causes of fish loss during May and June. No control is available for the treatment of bacterial problems in ponds. Fortunately, bacterial problems rarely reach epidemic proportions in ponds.

Vibriosis is also known as *Vibrio* hemorrhagic septicemia and is often associated with another disease, the red boil disease, which is caused by *Streptococcus* sp. The disease has been reported in *Epinephelus malabaricus, E. tauvina, E. coioides* and *E. bleekeri*. Vibriosis has been recorded in cultured groupers in Brunei Darussalam, Malaysia, Taiwan, Indonesia, Kuwait, Thailand, Singapore and the Philippines.

Causative Agents

The causative agents of vibriosis are *Vibrio parahaemolyticus, V. alginolyticus, V. vulnificus* and *V. carchariae*.

Stages Affected

The bacteria may affect grouper fry, fingerlings, juveniles, adults and broodstocks.

Gross Clinical Signs

The first sign of the disease is anorexia or loss of appetite with darkening of the fish coloration. The fish may be lethargic, swimming near the water surface. Affected fish may lose equilibrium and exhibit abnormal swimming behavior. One of the signs of the disease is body ulcer that may be hemorrhagic. Fin rot, which usually starts with erosion of the tip of the fin and gradually becomes necrotic, may also be observed. Exophthalmia and corneal opacity are also common signs of the disease. Internally, bloody discharges may be observed in the abdominal cavity due to internal organ hemorrhage. In the case of V. carchariae, gastroenteritis manifested by a swollen intestine containing yellow fluid may be observed.

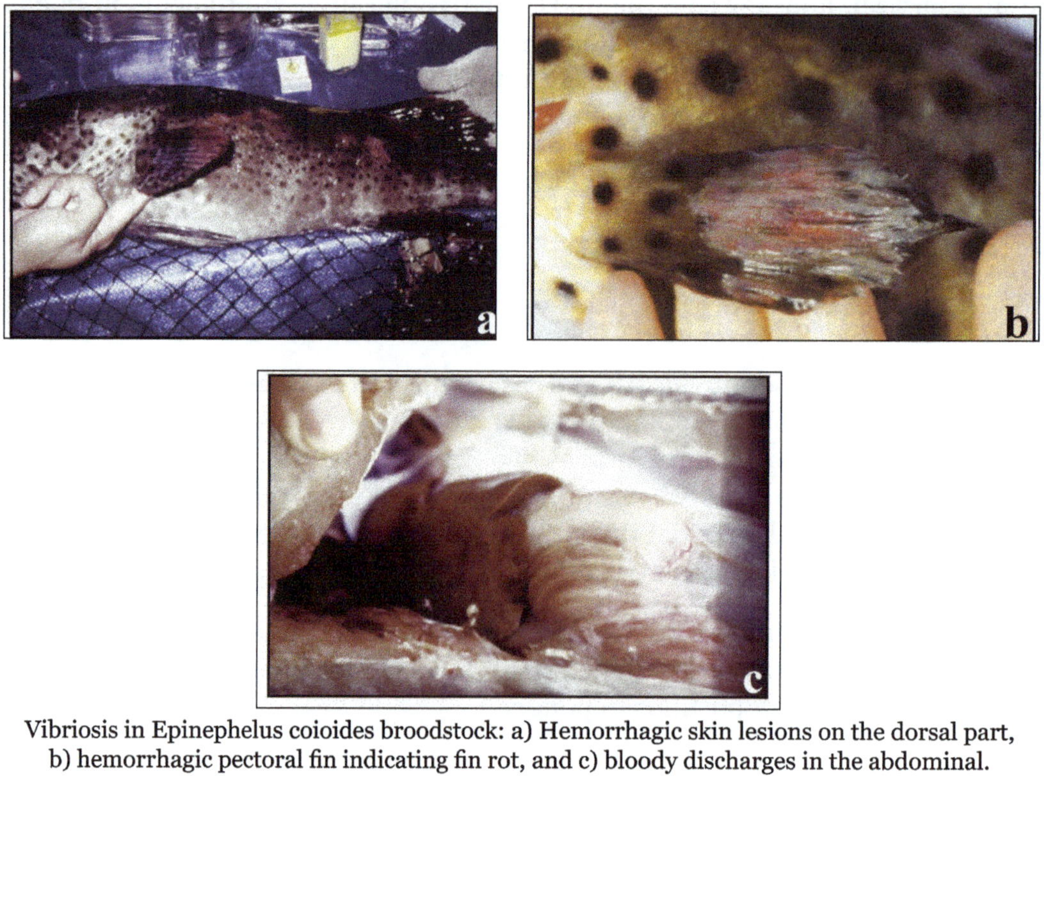

Vibriosis in Epinephelus coioides broodstock: a) Hemorrhagic skin lesions on the dorsal part, b) hemorrhagic pectoral fin indicating fin rot, and c) bloody discharges in the abdominal.

Effects on Host

Ten to 50% mortality may be observed in affected populations depending on the farm management. It has also been observed that infected grouper harbor monogeneans and protozoans on the gills and skin. Vibrio sp. infection is also associated with red boil disease. In relatively colder countries like Taiwan, outbreaks may occur in summer when ambient temperatures rise.

Transmission

The spread of the disease has been correlated with high salinity (30-35 ppt). The bacteria enter the fish through damaged areas. The disease may be transmitted through the water and trash fish given to the grouper. Infection is enhanced by parasitic infestation and mechanical injuries during transport, and grading.

Diagnosis

Squash preparation of affected areas examined under the microscope may reveal the bacteria. Vibrios are Gram-negative straight or curved rods, 0.5-0.8 μm in width and 1.4-2.6 μm in length. The bacteria may be isolated from the infected organ using tryptic soy agar (TSA), nutrient agar (NA), or brain heart infusion agar (BHIA) supplemented with NaCl. Thiosulfate citrate bilesalt sucrose agar (TCBS), a Vibrio selective medium, may be used to isolate and primarily identify associated main groups of vibrios. Opening of the abdominal cavity will reveal hemorrhagic internal organs. The kidney may be swollen and filled with yellow fluid.

Preventive Methods

Rough handling of the fish during stocking, sampling, changing of nets, grading and overcrowding should be avoided. Good water quality must be maintained.

Control Methods

The disease may be controlled through freshwater bath for 10-15 minutes. Affected fish may be treated with oxalinic acid mixed with feed at 20 mg/kg of fish. Terramycin added to feed at 7.5 g/kg for 5 days, reduced to 3.75 g/kg for the succeeding 5 days also proved effective. Prefuran bath treatment for 1 hour at 2 ppm may also be implemented.

Pseudomonas Infection

The disease is also known as pseudomonad hemorrhagic septicemia. The only reported case of the disease is in cage-cultured *Epinephelus tauvina* in Malaysia.

Causative Agent

The causative agent is *Pseudomonas* sp.

Stages Affected

The bacteria may affect grouper at all stages.

Gross Clinical Signs

Infected fish have extensive hemorrhagic erosions of the body. Ulcerations on the skin, fins and tails may also be observed. Other common signs of the disease are exophthalmia and corneal opacity.

Effects on Host

Twenty to 60% mortality may be observed in affected populations. A secondary epibiont fouling may also be observed. Internally, there is renal fragility, and dark red multifocal hepatic discoloration. Histologically, pathological changes consistent with subacute bacterial septicaemia are observed in the internal organs. Diffuse pericarditis and marked multifocal endocardial thrombosis and embolism are observed in the heart. Thrombosis and embolism are also observed in the hepatic vein. Diffuse pancreatic acinar cell atrophy and mononuclear cell infiltration are observed in the pancreas.

Transmission

Pseudomonas spp. are ubiquitous in the aquatic environment. Pseudomonas infects fish when it is subjected to environmental stressors such as extreme water temperature changes, overcrowding, poor water quality and sub-optimal nutrition.

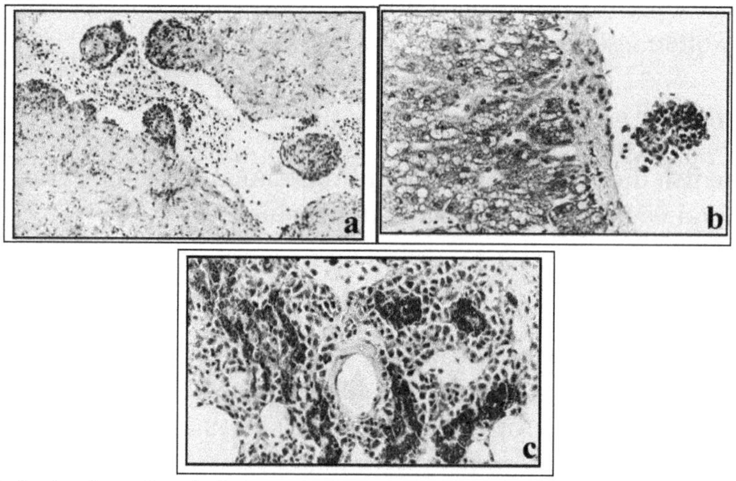

Pseudomonas sp. infection in Epinephelus tauvina: a) Endocardial thrombosis and embolism. H & E stain, b) embolism and early thrombosis formation in the hepatic vein. H & E stain, and c) diffuse pancreatic acinar cell atrophy and mononuclear cell infiltration.

Diagnosis

Squash preparation of the kidney and other internal organs as well as the affected areas may reveal large colonies of *Pseudomonas* sp. Pseudomonads are Gram-negative, straight or slightly curved rods but not helical in shape, 0.5-1.0 µm in width and 1.5-5.0 µm in length. The bacteria may be isolated from the infected organ and affected areas using glutamate starch phenol red agar (GSP) or *Pseudomonas-Aeromonas* selective agar, a medium that selectively promotes growth of *Aeromonas* and *Pseudomonas* but inhibits growth of other bacteria. Internal examination of affected fish will show renal fragility and dark red multifocal discoloration of the liver.

Preventive Methods

Avoid the predisposing factors such as extreme water temperature changes, overcrowding, poor water quality, and sub-optimal nutrition to prevent pseudomonad infection.

Control Methods

Transferring affected fish into another tank with clean water may control the infection.

Streptococcal Infection

The disease is also known as red boil disease. The disease is often associated with vibriosis. The disease has been reported in *Epinephelus malabaricus* and *E. bleekeri* in Brunei Darussalam, Malaysia, Singapore and Thailand.

Causative Agent

The causative agent is *Streptococcus* sp.

Stages Affected

Streptococcus sp. could infect grouper at all stages but is common in fry and fingerling stages.

Gross Clinical Signs

Affected fish are weak and display disoriented whirling motion. Exophthalmia and hemorrhages on the cornea, operculum, around the mouth and the anus could also be observed. Infected fish have red boils in the skin.

Effects on Host

The red boils on the skin enlarge and eventually burst, exposing the necrotic musculature underneath and form small ulcers, which act as portals of entry for other bacteria. The disease may also cause systemic infections with few external signs. The infection could cause 10% mortality in affected fry but is not fatal to older fish.

Transmission

The bacterium is ubiquitous in the environment, in the water and in carrier fish. Spread of the disease is associated with the presence of parasites, handling stress and sub-optimal water quality.

Diagnosis

The bacteria grow well on BHIA, Todd-Hewitt agar, horse agar and TSA supplemented with 0.5% glucose. The colonies on agar plates appear small (0.5-1.0 mm diameter), yellowish, translucent, rounded and slightly raised. *Streptococcus* spp. are Gram-positive bacteria with ovoid or spherical cells, less than 2 μm in diameter, occurring in pairs or chains.

Preventive Methods

Avoid the predisposing factors such as the presence of parasites, handling stress and suboptimal water quality to prevent disease occurrence.

Control Methods

Affected grouper could be treated with oxolinic acid mixed with feed at 20 mg/kg of fish, and perfuran bath for 1 hour at 2 ppm (see Appendix 1 for points to consider before using antibiotics and a list of anti-infectives recommended for use in marine food fish together with the withdrawal period).

Flexibacter Infection

Flexibacter spp. are long rod-shaped, Gram-negative bacteria with parallel sides and rounded ends, typically 0.5 μm wide and 1-3 μm long. The bacteria do not possess flagella and move by gliding, thus are also known as the gliding bacteria. Some *Flexibacter* spp. are widespread opportunistic bacterial pathogens. Some species of these yellow-pigmented bacteria have been associated with diseased fish, including *Flexibacter columnare* and *F. maritimus*, as well as other unidentified gliding bacteria referred to as *Cytophaga*-like bacteria. *Flexibacter* spp. are reported to cause bacterial gill disease and fin rot in groupers.

Bacterial Gill Disease

Bacterial gill disease has been reported in *Epinephelus malabaricus* and *E. bleekeri* in Brunei Darussalam and in *Plectopromus leopardus* in Indonesia. *Cytophaga* sp., *Flexibacter* sp. and *Flavobacterium* sp. cause bacterial gill disease in groupers in Brunei Darussalam. However, no bacteria were isolated from *Plectropomus leopardus* with bacterial gill disease in Indonesia, although histopathological examination showed rod shaped bacteria in the gills, which could possibly belong to the genus *Cytophaga, Flexibacter* or *Flavobacterium.*

Causative Agents

The disease is caused by *Cytophaga* sp., *Flexibacter* sp. or Flavobacterium sp.

Stages Affected

The bacteria usually attack fingerlings.

Gross Clinical Signs

Affected fish become anorexic, lethargic and dark in color. Fish tend to remain near the surface and may be flaring their operculum. The gills produce excessive amounts of mucus and the gill filaments may stick together. The gills of affected fish become yellowish in color indicating gill rot.

Effects on Host

A high mortality rate of >80% may be observed within a week in affected populations. The bacteria attach to the gill surface, grow in spreading patches and eventually cover individual gill filaments that result in cell death. Gill lesion may cause respiratory difficulty and the fish eventually dies. Histologically, fusion of the secondary lamellae, epithelial hyperplasia and presence of rod-shaped bacteria could be observed in the gills.

Transmission

The disease starts when the water quality deteriorates after a heavy rain. Silt and suspended organic particles from run-offs could irritate the gills and increase susceptibility to the disease. Low dissolved oxygen and high ammonia levels are often observed during disease outbreaks. Stress during grading makes fish susceptible to bacterial infection.

Diagnosis

The disease is diagnosed by the presence of brown to yellow brown growth of bacteria in the gills. Microscopic examination of wet mounts of the gill will reveal bacteria that are in a slow gliding movement. *Flexibacter* sp. is a thin, long Gram-negative rod with parallel sides and rounded ends, typically 0.5 µm wide and 1-3 µm long, that grows in layers, one on top of the other, giving it the appearance of "columns" or "haystacks" under the microscope.

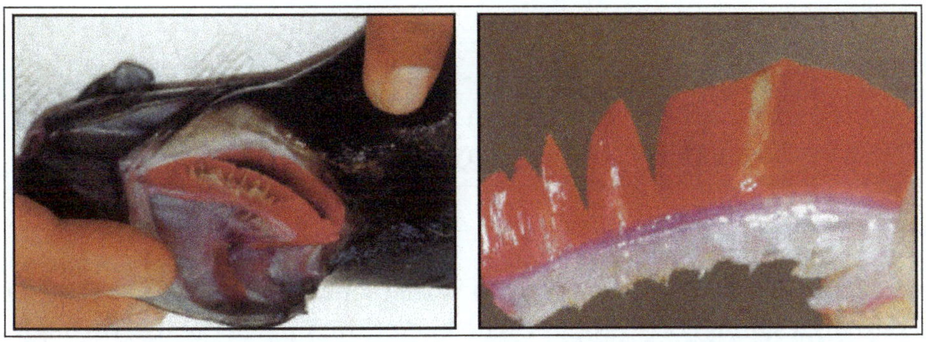

Bacterial gill disease in Plectropomus leopardus. Note the presence of brown to yellow brown growth of the bacteria in the gill filaments indicating gill rot.

The bacteria may be isolated using a selective media such as Selective *Cytophaga* agar and Hsu-Shotts medium. Colonies produced are pigmented yellow with rhizoids or root-like appearance.

Preventive Methods

Maintain good water quality and minimize stress by avoiding overcrowding, low dissolved oxygen and high ammonia levels. The occurrence of the disease could also be prevented by vaccination.

Control Methods

Transferring affected fish into another tank with clean water may control the infection. Affected fish may be treated with oxolinic acid mixed with feed at 20 mg/kg of fish and oxytetracycline at 75 mg/kg of fish/day for 10 days. Acriflavin dip at 100 ppm for 1 minute, and potassium permanganate at 2-4 ppm added to the water and allowed to dissipate over time could also be used to treat diseased fish (see Appendix 1 for points to consider before using antibiotics and a list of anti-infectives recommended for use in marine food fish together with the withdrawal period).

Fin Rot

Fin rot with hemorrhages usually affects *Cromileptes altivelis* from the wild in Indonesia. *Flexibacter maritimus* was isolated from the lesions. Secondary *Vibrio* infection may worsen the fish condition if not treated. Fin rot is also observed in fish infected with *Vibrio.*

Causative Agent

The disease is caused by *Flexibacter maritimus* (synonyms: *Cytophaga marina, Tenacibaculum maritimum*).

Stages Affected

The bacteria usually attack fingerlings.

Gross Clinical Signs

Affected fish become anorexic, lethargic and dark in color. Initially, the tip of the fin becomes grayish, and then it becomes eroded and hemorrhagic. The lesions progress into fin rot or extensive fin loss. Eventually, even the muscle fibers will be affected.

Effects on Host

Mortality rate of 80% may be observed within a few days if the infected fish are not treated. The bacteria could destroy the tail region within 2 days.

Transmission

The occurrence of the disease is correlated with water salinity. The disease is observed when the fish are exposed to high salinity of 30-35 ppt. The bacterium infects the fish through damaged area on the fin region.

Diagnosis

Squash preparation of affected areas examined under the microscope will reveal long rodshaped bacteria (0.5 × 2.5 µm) gliding slowly without flagella. The bacteria may be isolated from the infected tissue using *Cytophaga* agar prepared with seawater forming yellowish colonies.

Preventive Measures

Avoid rough handling of fish to minimize lesions, which could be portals of entry for the bacteria, due to mechanical damage.

Control Methods

Treatments should be implemented before secondary *Vibrio* infection sets in. Freshwater bath for 10-15 minutes and prefuran bath treatment at 1-2 ppm for 24 hours are effective in controlling the disease.

Cromileptes altivelis infected with *Flexibacter maritimus*: a) hemorrhagic and eroded caudal and pectoral fins and b) extensive loss of fin.

Fungal Diseases of Fish

Fish can suffer from various fungal diseases. Some of these diseases have been discussed here:

Exophialiasis

Exophiala salmonis and *E. psychrophila* are fungal organisms have hyphae that are septated, irregular in width and branched. Both fungal diseases infected the several species of fish. Symptoms of disease: 1. Fish become darker and lethargic, with erratic and irregular swimming behavior. 2. Round yellow to white granulomas are present in visceral organs like liver, kidney and spleen with prominent enlargement of the posterior kidney. Primarily it affects liver particularly but other severely affected organs are: spleen (in salmonids), heart (in herring), kidney (in salmonids), gonads, brain (in salmonids), gills (in salmonids) and musculature and nerve tissue behind the eyes (in sea fish).

Branchiomycosis (Gill Rot)

Branchiomycosis is to a large extent feared fungal disease of fishes approximately all over the world chiefly on carp farms. It is caused by the fungi Branchiomyces sanguinis (in carps) and Branchiomyces demigrans (in pike and tench). Branchiomyces demigrans differs from B. sanguinis by being able to proliferate from the blood vessels to the adjacent tissue of the gills and by having thicker-walled hyphae. The disease occurs most frequently in the warm climatic regions. Branchiomycosis is a pervasive problem in Europe, but has been only infrequently reported by US fish farms.

Branchiomycesis, another filamentous fungus obstructing the blood vessels in the gill filaments. Flecking on the gill filaments at a later stage becomes greyish-white and may at last drop off altogether leaving the cartilagenous support exposed. Both species of fungi are found in fish suffering from an ecological stress, such as low dissolved Oxygen, low pH (5.8-6.5) or a high algal bloom. The main sources of infection are the fungal spores carried in the water and detritus on pond bottoms. Branchiomyces sp. grows at temperatures between 57 °F and 95 °F, but grow best between 77 °F and 90 °F. B. sanguinis and B. demigrans infect the gill tissue of fish.

Gills appear striated or marbled with the pale areas representing infected and dying tissue. Fish may appear lackluster and may be seen gulping air at the water surface (or piping). Gills should be examined under a microscope by a trained diagnostician for authentication of the disease. Laboratory diagnosis of branchiomycosis is based on the examination of the gill filaments by the compression method.

Fungal hyphae are visible in the blood vessels of the gills at a 150-fold magnification under the microscope. Spoiled gill tissue with fungal hyphae and spores will be present. As the tissue dies and falls off, the spores are released into the water and transmitted to other fish. High mortalities are frequently linked with this infection. Spores can also be recognized when the diseases are in its acute stage. The affected tissues are torn into pieces on the microscope slide and examined by adding a drop of 50% glycerol solution in water or alcohol. The branchiomycetes are also well noticeable in preparations from dead fish bodies decayed by rot.

Saprolegniasis

This infection is caused by Saprolegnia parasitica. Initially, the fungus attacks the dead eggs and thereafter spread on the surrounding viable eggs resulting in their spoilage as well. This fungus often infects the fertilized eggs in hatching 'hapas'.

Dermocystidiosis

Among the most investigated marine fungal pathogen is Dermocystidium marinum. Researcher reported that warm-season mortality of oysters in the Chesapeake Bay was attributed to D. marinum. This high death rate was ascribed to increased environmental stress and warmer temperatures that aid proliferation of this fungus. It is important to note that Rhinosporidium seeberi, a close relative of Dermocystidium has recently been isolated from mucosal surfaces of human patients. reported a disease caused by R. seeberi, an aquatic protestant parasite producing rhinosporidiosis in human beings. Rhinosporidiosis manifests as slow-growing tumor like masses of the nasal mucosa or ocular conjunctiva in humans and animals.

Icthyophonus Disease (Swinging Disease)

Two species: Ichthyophonus hoferi and I. Gasterophilum is the responsible agent's for ichthyophonosis. Depending on the stage of the disease, different stages of growth of I. hoferican are distinguished microscopically: "resting" spore, germinating spore and hyphal stage. It grows in fresh and saltwater, in wild and cultured fish, but is limited to cool temperatures (36- 68 °F). The disease is transferred by fungal cysts, which are released in the faces and also by cannibalism of diseased fish.

The primary route of transmission is through the consumption of infective spores. Fish with a mild to moderate infection will show no external signs of the disease. In severe cases, the skin may have a 'sandpaper texture' caused by infection under the skin as well as in muscle tissue. Curvature of the spine may be found in some fish. Organs may be swollen with white to greywhite sores internally. The disease is called as "swinging disease" because diseased fish shows curious swinging movements. Researcher reported that Ichthyosporidium is a fungus, but it manifests itself internally. But it spreads in all organs but primarily attacks the kidneys and liver. Symptoms of disease include that fish may become sluggish, lose balance and eventually show external cysts or sores.

Treatment and Control of Fungal Diseases

Common treatments for fungal disease consist of use of potassium permanganate, formalin and provide iodine solutions. Bath treatment in NaOH (10-25 gm/L for 10-20 min), $KmNO_4$ (1 gm in 100 L of water for 30-90 min) or in $CUSO_4$ (5-10 gm in 100 L water for 10-30 min) should be given. Over treatment can promote damage of fish tissue, resulting in recurring infections. Environmental management is important for adequate resolution of chronic cases. If the disease is present, do not ship the infected fish. Averting is the best control for branchiomycosis. Good management practice will craft environmental circumstances unacceptable for development of fungi. Great care must be taken to avert movement of the disease to non-infected areas. Formalin and copper sulphate have been used to help stop mortalities; however, all tanks, raceways and aquaria must be sanitized and dried. Ponds should be dried and treated with quicklime (calcium oxide) and copper sulphate (2-3 kg/ha). An extended term bath in Acriflavine neutral or forma-green for 7 days helps in reduction of this condition. Dead fish should be buried. Saprolegniasis is best prohibited by good management practices. Sanitation should be observed and corrected when Saprolegnia is identified in an aquatic system. Good water quality and circulation, dodging of crowd to curtail injury (particularly during spawning) and good nutrition can prevent this condition. There is no cure for fish with Icthyophonus hoferi; they will carry the infection for life.

Cooking helps destroy the infective life stage. To avoid contamination of infective spores, never feed raw fish or raw fish products to cultured fish. If Icthyophonus disease is recognized by a qualified diagnostician, it is essential to eliminate and devastate any fish with the disease. 1% Phenoxethol solution added to food or Chloromycetin added to the food has also been effective against Icthyophonus disease. Ponds with dirt or gravel bottoms need months of drying to totally eliminate the fungus. Complete disinfection of tanks, raceways, or aquaria is encouraged.

Viral Diseases of Fish

Descriptions of viral diseases of fish are rapidly expanding. Viruses are being reported in new species, and interpretation of the significance of findings is also changing. Several viral diseases of ornamental fish are reportable.

Although viruses of homeothermic animals are cultured at uniform temperatures, fish viruses have wider, but specific, temperature tolerances in fish cell cultures at lower temperatures. Because of this relatively defined temperature range, variation in temperature may enable control, although often it merely induces latency. Because many viral diseases of fish are geographically limited, regulatory agencies and fish farms in disease-free areas consider them exotic diseases and require certification of introduced stocks. Many result in high mortality in young fish and little or no losses in adults, which may become carriers. For these reasons, avoidance of carriers and certification of SPF replacement stocks are frequently required. Specific testing procedures are available. Most vaccines used for control of fish diseases are for bacterial agents; however, use of vaccines to control some viral diseases is being introduced. Drugs are not effective; however, antibiotics and other drugs may be used to control secondary bacterial infections. Management techniques that minimize stress and crowding, biosecurity measures, and temperature manipulation hold the greatest promise for control of piscine viral diseases.

Herpesviruses

Carp Pox

One of the oldest recognized fish diseases, carp pox is caused by Cyprinid herpesvirus-1 (CyHV-1). Pox lesions may be seen on other species of fish and are sometimes referred to as fish pox. Lesions typically are smooth and raised and may have a milky appearance. They are benign, non-necrotizing areas of epidermal hyperplasia. Severe cases may result in development of papillomatous growths, which may be a site of complicating bacterial infection. Generally, lesions are self-limiting and of minimal clinical significance. Carp pox can be a significant problem in koi because their aesthetic quality, and hence market value, is severely compromised. For the serious koi enthusiast, fish affected with carp pox should be culled, preferably during quarantine. Surgical removal of the lesions has not been rewarding.

Koi Herpesvirus

Koi herpesvirus (KHV), caused by Cyprinid herpesvirus-3 (CyHV-3), was first recognized in 1996. It is widespread in the USA and considered endemic. Confirmed cases must be reported to the State Veterinarian and the USDA Area Veterinarian in Charge. Because the disease is endemic, regulating bodies report its occurrence to the OIE, but they do not require specific action, such as depopulation, when the disease is reported.

KHV causes clinical disease in koi and common carp. Goldfish and grass carp are refractory to clinical disease but may serve as carriers. Koi that are exposed, but survive, may also serve as carriers. Clinical disease is seen at water temperatures of 72°–81°F (22°–27 °C), with maximal mortality at temperatures of 72°–78 °F (22°–25.5 °C). Mortality rates can reach 80%–100%. Fish of any age are susceptible, but mortality rates may be higher in younger fish, especially fry. The most obvious lesions are seen on gill tissues, which are severely affected and develop a mottled red and white appearance with obvious hemorrhage in some cases. Affected fish are lethargic, swim at the surface, and may show behavioral signs of respiratory distress. The presence of severe bacterial or parasitic disease may mask the fact that KHV is the primary cause of gill lesions. The disease is transmitted horizontally by exposure to sick or carrier fish and also by exposure to contaminated water, substrate, or equipment.

When KHV is suspected, affected fish can be shipped to a laboratory for confirmation. Freshly dead specimens shipped on ice and received within 24 hr should be adequate. PCR or virus isolation and identification can be used to confirm the infection in dead animals. Because of the value of koi, there is significant demand for nonlethal testing protocols. Blood, gill tissue from biopsy, feces, or mucus may be used to assess the status of moribund fish. It is important to note these tests can result in false-negatives if performed on clinically normal fish. Indirect tests using blood samples, including ELISA or virus neutralization, are less straightforward. Negative test results could indicate a true negative or, alternatively, that the infection is in its early stages and the fish has not yet developed a measurable antibody response. A negative test result could potentially be obtained from a previously infected fish that has had a decrease in circulating antibody; however, such a fish could still function as a carrier. There is a misconception among some hobbyists that a negative blood test is an adequate screening test to exclude carrier status. Practitioners should communicate this clearly to owners, because it is easily misunderstood.

If KHV infection is confirmed in a population of koi, depopulation is strongly recommended. Surviving fish are carriers and can serve as a source of infection for naive fish. Although mortality decreases or stops as water temperature approaches 86 °F (30 °C), survivors will retain carrier status and therefore put other populations at risk.

Prevention of KHV is best accomplished with careful quarantine protocols. A minimum quarantine of 30 days at 75 °F (24 °C) is recommended to minimize the chance of introducing KHV to an established population of koi. If disease develops during the quarantine period, fish should be evaluated with special consideration for excluding KHV. After quarantine, a new fish may be placed in an isolated area with a few fish from the established population and monitored for signs of disease for at least 2 wk as an added precaution. If koi are taken to shows, quarantine protocols should be followed every time they return. Koi enthusiasts should be strongly encouraged to attend English-style shows (versus Japanese-style shows), where competing fish are not placed in a common container.

Herpesviral Hematopoietic Necrosis

This condition of goldfish is caused by Cyprinid herpesvirus-2 (CyHV-2). It is probably widespread throughout the USA, but it is not a reportable disease. Clinically ill goldfish often are anorectic and exhibit pale gills and ascites. At necropsy, the spleen and kidneys (anterior and posterior) are often enlarged. As is typical for herpesviruses, survivors can be carriers and exhibit clinical signs of disease if subjected to stressors. Water temperatures between 10° and 22 °C will result in replication of the virus that can be detected with quantitative PCR. There is no treatment.

Herpesvirus of Angelfish

A herpesvirus of angelfish (*Pterophyllum* spp) has been detected by electron microscopy of skin from moribund angelfish. Affected fish produce copious amounts of skin mucus that gives affected fish a grey sheen. Often, these fish have multiple parasitic infestations and bacterial infections, similar to that of KHV in koi. It is suspected that survivors are carriers.

Rhabdoviruses

Viral Hemorrhagic Septicemia (VHS)

This reportable disease is caused by a *Novirhabdovirus* and is a member of the family Rhabdoviridae. This disease is listed by OIE as notifiable. Most of the reported hosts are not ornamental fishes, but koi were shown to be susceptible experimentally to genotype IVb. This genotype affects a diverse group of fishes in the Great Lakes region of North America.

Spring Viremia of Carp (SVC)

This acute, virulent, usually hemorrhagic disease of cultured carp is caused by a *Vesiculovirus* that, like VHS, is a member of the Rhabdoviridae family. The disease is listed as notifiable by the OIE. Historically, it was reported in Europe and the former USSR; however, several outbreaks have been reported in the USA between 2002–2007, in both wild fish and cultured ornamental koi. SVC is considered a foreign animal disease in the USA and must be reported. It causes disease in common carp, including koi, as well as grass, bighead, silver, and crucian carp. Limited experience suggests that common goldfish may be susceptible.

Clinical signs are nonspecific and may include darkening of the skin, exophthalmia, ascites, pale gills, hemorrhage, and a protruding vent with thick mucoid fecal casts. Pinpoint hemorrhage in the swim bladder is indicative of SVC, if present. Coinfection with *Aeromonas* or other systemic bacteria may obscure the presence of the virus. The bacterial component of the infection can be controlled with antibiotics; however, depopulation of affected or exposed fish is required in the USA. The disease causes death in both adult and young fish. Clinical disease occurs at cool temperatures, 54°–72 °F (12°–22 °C), an important distinction from KHV. The virus is readily isolated in common fish cell lines and identified by serum neutralization and fluorescent antibody tests.

Iridoviruses

Lymphocystis Disease

This typically chronic, viral infection of wild or captive marine and freshwater fish is caused by an icosahedral DNA virus of the Iridoviridae family. Infection may be manifest by benign, cauliflower-like lesions typically located on fins. The disease affects a wide range of fish and is generally considered to have a global distribution. Within the aquarium trade, painted glass fish and marine tropical fishes such as the anemonefishes (Pomacentridae), marine angels (Pomacanthidae), and butterflyfishes (Chaetodontidae) are susceptible. Presumptive diagnosis is based on the presence of enlarged fibroblasts (up to 1 mm), which are easily visualized with a light microscope. Microscopic examination typically reveals the appearance of grape-like clusters of virus-laden cells. Diagnosis is confirmed histologically. Feulgen-positive cytoplasmic inclusions and a hypertrophied nucleus are pathognomonic. The disease is usually self-limiting but is of aesthetic concern.

Megalocytiviruses

Several other iridoviruses have been described in ornamental fish. Megalocytiviruses consist of two major groupings, red sea bream iridovirus (RSIV) and infectious spleen and kidney necrosis virus (ISKNV). RSIV has been found in >30 other marine fishes so far; many of the affected species include some tropical fishes kept in aquaria. RSIV is a reportable disease. ISKNV includes megalocytiviruses that primarily affect ornamental freshwater fishes. ISKNV is not currently reportable.

A megalocytivirus has been described in freshwater angelfish (*Pterophyllum scalare*), but the agent has not been isolated. It can result in high mortality. An iridovirus has also been isolated in the USA from gourami in the genus *Trichogaster* using a tilapia heart cell line. This virus does not grow on FHM cells or other common cell lines used for isolation of fish viruses. The gourami virus has been associated with systemic disease and mortality of *T trichopterus* and *T leeri* gourami. Efforts to fulfill River's postulates were supportive although not conclusive. Clinical disease with the USA gourami megalocytivirus appears to be more severe at water temperatures ≥30 °C, based on very limited information. A megalocytivirus is known from dwarf gourami *Trichogaster* spp, but it is not known whether it is the same virus that affects gourami in the USA.

Other Viruses

Koi sleepy disease is also known as carp edema virus. Caused by a poxvirus, it was first seen in Japan in 1974 and has since occurred in other countries, including the USA. Affected fishes appear lethargic, lay on their sides, and have thin body condition. Mortality is usually low but can be variable. Gill hyperplasia is a common histopathologic finding. Infected koi can recover, but it is unknown whether they retain a carrier state.

Fish Parasites and its Diseases

Parasites have a wide range of distribution in all groups of animals. They are more abundant than free-living animals, and may be found in every phylum of animal from protistan to chordates. A large number and diversity of animal species are capable of patrasitising fish, ranging from microscopic protozoans to easily visible crustaceans and annelids. Most of the fishes, either wild or cultured, are infected with parasites. They not only serve as the host to different parasites, but also serve as carriers of many larval parasitic forms that mature and cause serious diseases in many vertebrates, including man. Parasites exhibit marvelous strategies for adaptation to their hosts. Many parasite species are host-specific to some extent and are capable of infecting one or only a limited number of host species. Individual parasite species may also have widely differing effects on different host species. Some of them are parasitic in the external surface of fish; others are parasitic in the internal organs. They can infect fishes in different stages of their life, as well as different aquatic environmental conditions, and are also considered to be biological indicators of environmental pollution. Parasites interfere with the nutrition, metabolism, and secretory function of the alimentary canal, damage nervous system, and also upset the normal reproduction of the host.

Most parasite species rarely cause problems in the natural environment but in aquaculture, parasites often cause serious outbreaks of disease. They play an important role in determining the productivity, sustainability, and economic viability of aquaculture. Parasite infections cause serious socioeconomic, ecological, and welfare consequences in global finfish aquaculture. They are responsible for direct mortality of farmed stock, retarding growth rate and feed conversion ratios, treatment costs incurred, and rejection of product during processing. Besides direct losses, parasites may have considerable impact on the behavior of fish, reduced fecundity, their resistance to other stressing factors, susceptibility to other infections, the potential legislative burdens, and their presence may also reduce marketability of fish. Parasites generate costs for managing infections as well as prophylactic treatment in finfish aquaculture. The annual global cost of parasites in finfish aquaculture can be tentatively estimated at $1.05 billion to $9.58 billion. Moreover, many of the parasites are also of zoonotic importance, and have public health consequences. Without stringent and appropriate control measures, the impacts of these pathogens can often be significant. It is therefore an important area for proper attention to be given by the scientists for sustainable and safe fish production. This chapter presents an overview on the important parasitic infections in fish.

Protozoan Diseases

Most of the commonly encountered fish parasites are protozoans. They are single-celled organisms, many of which are free-living in the aquatic environment. Their ability to multiply on or within their hosts makes them in many instances very dangerous to fish. They have a direct life cycle and mostly reproduce by binary fission; some species have cyst form, off the host. Typically, these parasites are present in large numbers either on the surface of the fish, within the gills, or both. The general effect of these parasites is to irritate the epithelial surface, causing an increasing mucus production. There are three main groups of protozoan parasitising the external tissues of fish: ciliates, flagellates, and amoebae.

Ciliates

Protozoa belonging to the Ciliophora are equipped with cilia (short, fine cytoplasmic outgrowth), or a structure derived from cilia by secondary modifications, or both. Ciliated protozoa are among the most common external parasites which cause mortalities in a number of wild and farmed fishes. Cilitates can be motile, attached, or found within the epithelium. While they often occur as harmless ectocommensals, under poor environmental conditions or stress, some ciliates can rapidly increase in number, leading to morbidity and mortality.

Ichthyophthiriasis

Ichthyophthirius multifillis, referred to as 'Ich' is the causative agent of Ichthyophthiriasis or white spot disease. It is the most common pathogen of protozoan parasites of freshwater fishes worldwide. It causes particularly devastating infections in farm-raised fish, where it spreads rapidly within dense populations, leading to extensive morbidity and mortality. Ich infection can occur at any of the growth stages of fish, from fry, fingerling, table size, to brood fish. The parasite can cause catastrophic epizootics in warm and temperate-water fish culture, and may even cause losses in wild fish on occasion. Considerable losses have been reported from cultures of carp, rainbow trout, tilapia, eel, channel catfish, as well as ornamental fish. Ich infections in fingerlings affected about 4% of total channel catfish producers in the U.S. in 2009. In Europe, annual losses to trout farmers are estimated to be approximately $140 million per year. Ich epizootics were reported in China as early as 10th century. The first major outbreak in North America was described in 1898. From 1940, it became a serious disease of carp in Russia. The disease has been reported in fry and fingerlings of Indian major carps from nursery and rearing ponds in Bangladesh.

Ichthyophthirius multifiliis is spherical in shape and the cilia are evenly distributed over the whole surface. Characteristic feature of the parasite are the horse-shoe shaped nucleus, and its rotating movement. It is an obligate pathogen and has a unique direct life cycle which allows a rapid intensification of infection. The life cycle consists of an infective theront, a parasitic trophont, and a reproductive tomont. Infective theronts swim actively in water in search of hosts. After burrowing into fish epithelium, theronts become trophonts and feed on host tissue until they reach maturity. The mature tomont drops off the host, attaches to substrates, and undergoes multiple divisions to produce theronts. The parasite spreads rapidly from fish to fish as a single Ich tomont can produce hundreds to thousands of infective theronts in less than a day.

The parasite invades epithelial tissue of gills, skin, or fins, leaving a small wound and visible white spot or nodule where each parasite encysts. Infected fish are extremely lethargic and covered with visible white dots. Mortality can be rapid and catastrophic. Heavy infection by Ich damages fish skin and gills, causes loss of the respiratory, excretory, and osmoregulatory functions, and might serve as a portal of entry for secondary invaders, leading eventually to death of fish.

Trichodiniasis

Trichodiniasis is one of the major protozoan diseases found in fish worldwide. The disease is caused by a large assemblage of peritrichous ciliates of trichodinids. The trichodinid group includes *Trichodina, Trichodinella*, and *Tripartiella*, which are important ectoparasites of freshwater and marine fish worldwide. They are capable, in some cases, of causing heavy damage to their hosts, resulting in mortalities. Infestations caused by trichodinids are particularly significant in aquaculture because they are responsible for causing decreased growth, chronic mortality during cage production and changes in vision and swimming in larvae, culminating in acute mortality. Trichodinids have a monoxenic life cycle, and reproduce mainly by binary fission on the host. They can be horizontally transmitted by direct contact or by contaminated water, in which the parasite searches for new hosts. Outbreaks are often associated with poor water quality and stress. Contaminated fish farming utensils are also another important source of transmission of trichodinids. Trichodinids are up to 100 μm in diameter, and have a basic saucer-shape with a fringe of cilia around the perimeter which are used for locomotion and feeding. Their body is supported by a rigid ring of interconnected discs called a chitinoid or denticular ring. The parasite brows over the surface of gills and skin with spinning motion, damaging the host tissues and consuming the resulting tissue debris. Infected fish often have a grayish sheen due to excess mucus production, and fish may become frayed. Ultimately, erosion of the epithelium will occur. The main pathological changes associated with this parasitic infection are related to gill tissue, such as hyperplasia, hypertrophy, oedema, inflammatory infiltration, and necrosis.

Chilodonellosis

Chilodonella is a highly pathogenic holotrich ciliate, ectoparasite on the skin and gill of a wide range of temperate and tropical freshwater fish. The parasite has a flattened, ovoid shape, is up to 80 μm in length, and is covered by rows of cilia which move it in a steady gliding manner over the epithelial cells of its fish host. Heavy infections of *Chilodonella* are often associated with poor water quality. Carps, salmonids, and catfish are the species most commonly affected. *Chilodonella hexasticha* is most likely to be problematic at lower water temperatures, and is reported as a serious pathogen in overwintered carp. *Chilodonella piscicola* (=C. *cyprini*) infects cyprinids particularly, but can be found on other fish, where it can cause problems at higher temperatures. Fingerlings can be especially vulnerable.

The mass development of parasite causes a much higher production of mucus and disturbances in the respiratory function of the skin. The fish is restless, and rises to the upper layers of the water. Its entire body is covered with the bluish-white coating, and is particularly noticeable in the head region. A smear in the skin surface reveals numerous individuals of the parasites. Damage to the skin may then be subject to secondary invasion by bacteria, fungi, and other pathogens. It causes localized hyperplasia of the gill epithelium. The thin respiratory epithelium is covered by the hyperplastic epithelium, and this drastically reduces the respiratory surface of the gills.

Flagellates

Flagellated protozoans are small parasites that can infect fish externally and internally. They are characterized by one or more flagella that cause the parasite to move in a whip-like or jerky motion. Flagella are longer and more powerful than cilia and are always few in numbers.

Ichthyobodosis

The disease is caused by heavy infections on skin and gills of fish by parasitic flagellates belonging to the genus *Ichthyobodo*. It is an important disease that has caused severe loss among farmed and ornamental fish worldwide for more than a century. *Ichthyobodo* is regarded as one of the most damaging parasites among farmed salmon, and is probably the major cause of mortality among salmonid fry and fingerlings. The flagellate that is perhaps best known as a serious fish pathogen is *Ichthyobodo necatrix* (=*Costia necator*), which causes the disease known as *costiasis*.

Ichthyobodo spp. spread rapidly between hosts in fish farms, most likely by both direct contact or through free-swimming parasites. Heavy infections may occur when conditions favor the parasites. Poor rearing conditions such as low water flow and high crowding densities are considered particularly important. Massive infections on skin and gills can cause epithelial hyperplasia or hypertrophy, and may result in severe or fatal osmoregulatory or respiratory problems. Clinical signs of Ichthyobodo spp. are an excess mucus production that forms a blue-grey or white film over the body surface and gills. There have also been described several non-specific clinical signs of severe and prolonged *Ichthyobodo* spp. infections, including flashing, lethargic behavior, listlessness, loss of appetite, and increased mortality.

Ichthyobodo spp. are obligate ectoparasites with a direct life-cycle on the host, and they cannot subsist or multiply without an appropriate host. There are two forms: a parasitic feeding form which attaches to the fish's epithelial cells and a non-feeding, swimming form that exists off the fish. The free-swimming form is kidney-shaped with two pairs of flagella. The attached form is pear-shaped and attaches to the gill and skin. The parasites disappear from a dead host, and die after 30–60 minutes in the free-swimming form outside a host.

Hexamitasis

Hexamitiasis is caused by excessive numbers of flagellated protozoa of the genus *Hexamita* in the alimentary tract of farmed and wild freshwater fishes. *Hexamitia* sp. is a small (10 µm) pear-shaped, pyriform organism with three anterior and one posterior pair of flagella. *Hexamita truttae* is common in North American trout hatcheries, which cause mass mortality of fish. Clinically, the young fish have anorexia, debilitated with reduced growth, have trailing faecal casts, excessive nervousness, and the abdomen may be distended. The fish develop an acute enteritis, yellowish watery gut contents with numerous organisms present in the faeces or bile from the gall bladder. Transmission is by ingestion of an infective cyst. In farmed Chinook and Atlantic salmon, the disease can become systemic with fish becoming anaemic with swollen kidneys and exophthalmus. *Hexamita salmonis* may be present in vast numbers in the pyloric intestine of cultured salmonids in freshwater, but the degree of damage it causes has been disputed. These parasites present a monoxenic life cycle; the pear-shaped trophozoites change to spherical before cellular division. They can be horizontally transmitted by the releasing of trophozoites and oocysts into the water from the fish feces that will be ingested by other hosts.

Amoebae

Some free-living amoebae may change their mode of life and become harmful. Several species of amoeba have been implicated in gill disease in salmonids.

Amoebic Gill Disease

Amoebic gill disease (AGD), caused by the free-living, facultative amoeba *Paramoeba* [=*Neoparamoeba*] *perurans*, is a major issue in marine salmon farming, which leads to gill damage and death of infected fish. It has become a significant problem in sea-caged Atlantic salmon and rainbow trout in Tasmania, and has been considered the most serious infectious disease. Cases of gill amoebic infections of fi other than salmonids have also been reported, e.g., in European catfish (*Silurus glanis*) or turbot. Clinical AGD most often occurs at water temperatures of 10–20 °C, and is sometimes associated with higher than normal temperatures. Gross pathology in infected fish is characterised by raised, multifocal, white mucoid patches on the gills, which represent regions of epithelial hyperplasia of the primary and secondary lamellae. This phase is followed by desquamation of the epithelium, local disturbances of blood circulation, and progressive changes represented by inflammation. All the above mentioned changes result in decrease or loss of gill respiratory surface area. Significant cardiac changes and acid-base disturbances may occur in AGD affected fish, which may result in acute cardiac dysfunction and death.

Metazoan Diseases

Metazoan parasites include the myxozoans, helminths, crustacean, annelids, and mollusks are common in both wild and cultured fish. All have direct life cycle, but the period of their life cycle in which they act as fish parasites varies considerably with the species concerned.

Whirling Disease

Whirling disease is a chronic inflammatory disease in salmonid fish caused by the myxozoan parasite *Myxobolus cerebralis*, which is characterized primarily by the tight circular movements due to spinal cord constriction and brain stem compression in infected fish. It occurs worldwide and rainbow trout are most susceptible to disease, although the parasite is able to infect numerous species of salmonid fishes. It has been suggested that *M. cerebralis* was originally a parasite of relatively low pathogenicity of native salmonids, such as the brown trout and Atlantic salmon in central Europe. The introduced rainbow trout, on the other hand, is highly susceptible to whirling disease. Through shipments of live and frozen trout, perhaps even with contaminated trout ova, *M. cerebralis* became established in most countries where rainbow trout are cultured. It is now a serious and expanding problem in the native wild populations of rainbow trout and cut-throat trout in the western United States.

Spores of the parasite are oval and have two distinct polar capsules. This protozoan has a complex life cycle. Spores can be shed from infected live fish, as well as from dead and decomposing fish. The spores can also be spread via bird feces. Spores are ingested by an annelid worm intermediate host, *Tubifex tubifex*, which lives in the bottom mud of ponds, streams, and earthen raceways. The spores develop into actinosporeans that penetrate fish (or the fish ingest the actinosporeans when they eat tubifex worms). Plasmodia develop in the fi cartilage and eventually produce the characteristic spores.

The disease attacks the cartilage. The development and severity of whirling disease depend on the age and size of the salmonid host when exposed to the parasite. Very young fish are most vulnerable to *M. cerebralis*, and susceptibility decreases with age and growth, as bone replaces cartilage in the developing fish. Heavy infections in young fish often result in death. *Myxobolus cerebralis* infections in the spine can cause the fish's tail to turn black and the spine to curve. A blackened tail is caused by pressure on nerves that control pigmentation. Infections in the head cartilage create head and jaw deformities, while infections in the auditory capsule cause young trout to become disoriented and chase their tails in a whirling motion. Permanent deformities of the head, spine, and operculum are caused by cartilage damage, associated inflammation, and interference with normal bone development. During the active phase of infection growth rates may be depressed.

Proliferative Kidney Disease

Proliferative kidney disease (PKD) caused by the malacosporean parasite *Tetracapsuloides bryosalmonae,* is a widespread disease that causes significant losses among farmed salmonids in Western Europe and North America. PKD is also the suspected cause for several declines in wild salmonid populations. The disease is characterized by a chronic inflammation of the anterior and posterior kidney, which is caused by a massive accumulation of lymphocytes and extensive granuloma formation.

Fish are infected by parasite spores released from bryozoans, which are the invertebrate hosts of the parasite. The parasite invades the fish through skin and gills, and afterwards invades inner organs, with the kidney being the main target organ for further development. Infection and clinical signs of PKD are dependent on environmental temperature. Water temperatures above 12 °C may induce clinical PKD, but normally, clinical PKD outbreaks occur at water temperatures above 15 °C. It primarily affects fingerling fish, and usually results in 100% morbidity on an affected farm, with up to 20% mortality.

Diseases Caused by Crustaceans

A tremendous number of crustaceans have evolved to become dependent on certain animals for existence. Those closely associated with fishes can cause disease problems. Parasitic crustaceans are among the most serious gill and skin parasites of fish worldwide. They may be found attached to the external surfaces of both marine and freshwater fish.

Argulosis

Argulids, commonly referred to as "fish lice", are obligatory ectoparasites causing problems in fisheries and aquaculture worldwide. About 150 Argulus species are known from marine, brackish, and freshwater habitats. *Argulus foliaceus* L. 1758, is probably the most common and widespread argulid in Western Europe. This species has been reported as a threat to the culture of tilapia, rainbow trout, and common carp. *Argulus japonicus* Thiele, 1900, is believed to have originated in Japan, and probably owes wide-spread distribution to the trade in ornamental fish, such as goldfish and koi carp. Argulosis represents a major problem within the Bangladeshi carp culture industry, with infections causing mortality, morbidity, and growth loss.

At 5–10 mm, *Argulus* is a large parasite that is visible with the naked eye. Argulids have a very distinctive oval-shaped, flattened carapace. Other notable physical features include compound eyes, a pair of large suckers, four pairs of branched thoracic swimming limbs, and a small unsegmented abdomen. While most fish lice are effective swimmers, many species tend to move through the water in a loose cork-screw motion or a somersaulting action. *Argulus* has a four stage life-cycle (egg, nauplius, meta-nauplius, adult), and requires a fish host at least once during their life. The parasites complete their life-cycle within 30–100 days, depending on water temperature. They are transmitted via free-swimming meta-nauplii stages. Once emerged from eggs, non-feeding meta-nauplii actively search for a host. Without a host fish, lice survive only for a few days, and the juveniles for an even shorter period (< 48 hours).

The lice attach themselves to the skin of the host with their strong suckers, and then feed on its host's blood or mucus by using a modified disk which possesses piercing and sucking mouthparts. During feeding, argulids inject cytolytic enzymes and then draw off blood. Localised inflammation is often seen at the site. In the early stages of infection, an increase in the frequency of fish jumping and a reduction in feeding have been observed. As the infection progresses, secondary fungal and bacterial infections develop and fish may exhibit shoaling behavior. If the infection continues, there may be large-scale mortalities. In addition to their destructive nature, *Argulus* species are known to act as vectors for other pathogens including viruses, bacteria, fungi and nematodes.

Lernaeasis

Lernaenid copepodes or 'Anchor worms' are important ectoparasite parasites in freshwater aquaculture of cyprinids, and occasionally of salmonids and other fishes. Epizootics in cultured fish are often associated with high mortality. The parasites also cause problems in commercial aquariums. Infection by this parasite has been associated with reduced host weight, growth, and fecundity. The very common *L. cyprinacea* is considered one of the most invasive ones, having worldwide distribution. A number of species have been reported from Indian region (*L. chackoensis and L. bengalensis*). Indian major carps are particularly affected by lernaeasis in Bangladesh. Infestations with Lernaea are most prevalent in the summer months and occur more commonly in stagnant or slow-moving water bodies.

Only the female stage is parasitic and is found embedded in the dermis of the fish sometimes penetrating the musculature and reaching the peritoneal cavity in small fish. Attachment is by means of a branched anchor formed by modification of the head region. The embedded parasite induces a necrotic ulcer around its insertion, and eventually a connective tissue capsule forms around the head of the parasite. Lernaea can cause intense inflammation, leading to secondary bacterial and fungal infections. These secondary infections sometimes worsen and kill the fish. Larger numbers of parasites on the gill can interfere with respiration, causing death. Fish can survive *Lernaea* infection, but chronic conditions frequently result in poor growth and body condition. Mortalities occur most often in young fish, but even if they survive they lose weight and are unsightly.

Disease Caused by Monogenetic Trematode

Monogenetic trematodes are parasitic flatworms or flukes that generally live on the surface of a fish host. Damage is caused to the host by the penetration of their attachment organ and by browsing action of the mouth at the free end. They have been recognized as a serious pathogen of fish in aquaculture. Monogeneans are able to multiply rapidly in high density aquaculture environments because they have a direct single host life cycle, as they require no intermediate host. They produce freely deposited eggs that often become entangled to high re-infection rates amongst fish. Severely affected fish may die, as a result of gill pathology and interference with the exchange of respiratory gasses and ions. Monogeneans display a high degree of host specificity. They can be viviparous or oviparous; are hermaphrodite and commonly protandrous. They attach to their host by means of a specialized muscular posterior attachment organ or opisthaptor, which houses a variable array of sclerotized hooks, connecting bars, clamps, or epidermal structures. Two most common representatives of this group are *Gyrodactylus* and *Dactylogyrus*.

Gyrodactylosis

The disease is caused by the genus *Gyrodactylus*. They are commonly known as 'skin flukes' and are found all over the surface of the body, including fins and occasionally in the gills of both marine and freshwater fish. *Gyrodactylus salaris* is perhaps the most economically impactful ecto-parasitic flatworm of fish which has been responsible for mass mortalities of wild Atlantic salmon when transferred to Norwegian rivers. It is the only notifiable flatworm disease listed by the Office International des Epizooties (OIE) (World Organization for Animal Health). Most species of gyrodactylids are very host-specific and may be even be specific to a particular site on the host.

Gyrodactylids are small worms, about 0.3–1.0 mm in length, and may just be visible to the naked eyes. Adult parasites carry a fully developed embryo identical to the adult, which in turn carry young of next generation. They have a pair of anchors with both dorsal and ventral bars and 16 marginal hooks, and do not have eye spots. Attachment to the fish is made with the marginal hooks; the anchors are used as a spring-like device to assist attachment with the marginal hooks. An embryo with its pair of anchors may frequently be seen inside an adult gyrodactylid. As they are viviparous, gyrodactylids are able to reproduce extremely rapidly if conditions are favorable. Transmission is thought to be by direct fish to fish contact, although parasites may survive for some time in the water column if detached from their host. They move by a characteristic looping action, and small puncture wounds occur each time the hooks penetrate the host. Heavily infected fish may have increased mucus production, frayed fins, skin ulcers, and damaged gills. The lesions are caused by the feeding activity of the parasites. The occurrence of gyrodactylids in epizootic proportions in cultured fish is generally a sign of poor environmental conditions and stressed fish.

Dactylogyrosis

Dactylogyrosis is caused by *Dactylogyrus* species, and in general they are known as "Gill fluke". Some species of *Dactylogyrus* have proved to be very dangerous in cyprinid culture, especially to fry, where relatively few parasites can cause severe gill damage. The intensity of infestation rises rapidly to reach several hundred per fish. Dactylogyrid flukes are common in freshwater fishes in Bangladesh. A numbers of species of this genus (*Dactylogyrus mrigali, D. chauhanus, D. yogendrai, D. labei, D. kalyanensis*) have been reported from the Indian major carps and from other several exotic and indigenous fishes.

Dactylogyrids have two to four eyespots, one pair of large anchor hooks, and are egg layers. They are oviparous and relatively small. Anteriorly there is a pair of pigmented light receptors and two cephalic lobes, which contains secretions of adhesive gland cells. They have a muscular pharynx and a tubular, posteriorly confluent intestine. The haptor has two pairs of hamuli and 14 marginal hooks. The eggs hatch into free-swimming larvae and are carried to a new host by water currents and their own ciliated movement. Eggs are laid by dactylogyrids embryonate and hatch in a period which varies greatly with temperature within a species and between species. In some species, in temperate climates, eggs laid in autumn may overwinter and hatch on the return of favorable conditions in the spring, when newly hatched fry of host species are available. Dactylogyrids also have optimal temperatures for egg production, resulting in seasonal peaks of infection, and also different geographical distributions between parasite species.

Clinical signs of fish infected by gill fluke include increased breathing frequency, whilst the gill coverings are stretched open widely, the gill being expanded and very pale. Parts of the gills often become protuberant and show as a small pale fleece outside the covering. Parts of the gill sheets on which flukes have settled are covered with a cloudy film, consisting of slime and destroyed epithelial cells. The gills often suffer from heavy damage. Hyperplastic changes of gill epithelium often spread to areas not colonized by the worm. Telangiectasis (gill blood blisters) is frequent and widespread. Local tissue erosion at the attachment site is accompanied by vigorous peripheral proliferation. Mucus production is over stimulated, seriously impeding respiration. Gills are discolored and swollen so much that normal closure of the opercula may be impossible.

Diseases Caused by Digenetic Trematode

Digeneans are endoparasitic flatworms, also known as 'flukes' or 'digenetic trematodes', principally infecting the alimentary canal or associated organs. The principal developmental stages of digeneans are called miracidium, sporocyst, redia, cercaria, metacercaria, transitory migrating larva, and adult. They have a complicated indirect life cycle with at least one intermediate host. The stage most commonly encountered in fish is the larval metacercaria, which may be found in the tissue within the cyst or unencysted, depending on species. The miracidium is generally ovoid and possesses cilia covering the body that allow the parasite to swim and search for the first intermediate host, which is usually represented by a snail. The final host is usually a piscivorous bird. Damage can occur to the fish when cercarial larvae first invade through the skin of the fish. Some metacercaria are extremely damaging to their target organ, while others cause very little harm, although their presence may be aesthetically unappealing.

Diplostomiasis

Diplostomum spathaceum referred to as 'eye fluke' is the causative agent of diplostomiasis. The parasite invades the lens of many species of fresh- water fish in Europe and North America. The infection of eye fluke can cause cataract and blindness in a range of fish species, of which Cyprinids and rainbow trout appear to be particularly susceptible to *D. spathaceum*. The first intermediate hosts are *Lymnaea* spp. and the final hosts are primarily gulls (*Larus* spp.). The metacercariae in the lens cause cataract formation and eventually blindness so that the fish is unable to feed and loses condition. It also becomes more susceptible to predation by piscivorous birds, in which adult parasite is found. The numbers of metacercariae required to cause blindness obviously vary according to the size of the fish. Although the fish is not killed by blinding, its growth rate may be greatly reduced due to its inability to feed normally, and such fish may become emaciated.

Fish Leech

Leeches are clitellate annelids and are the only important fish pathogens in the phylum. Aquatic leeches, both freshwater and marine, occur in a diversity of habitats worldwide. Leeches are segmented worms that feed on blood. Armed with suckers at each end of the body, they are able to loop over the body of the fish to feed. Leeches can potentially affect the health of fishes in a variety of ways, but most involve blood-feeding activities. In freshwater, *Piscicola geometra* can reach epizootic levels in rainbow trout and cyprinids cultured in earthen ponds. In the marine environment *Hemibdella* sp. has been noted as a problem on cultured Dover sole and turbot. Some species, however, can on occasion become important fish ectoparasites. Leeches act as vectors or intermediate hosts of protozoan parasites among like *Cryptobia*.

The fish attacked by a leech shows, especially during the early stages of attachment, signs of irritation and restless. It may attempt to dislodge the parasite by rubbing against objects. Its movements may become unsteady and erratic. Moribund conditions caused by leeches are easily identifiable because they are accompanied by the presence of large numbers of these parasites. Secondary effects of infections might be more serious than the pathogenic changes. Leeches by themselves are only rarely associated with serious pathological effects. Generally the lesions associated with their blood meals are clean and heal readily once the bleeding stops. Mortalities are rare, and usually the effects of the leeches are restricted to growth reduction.

Fishborne Zoonotic Diseases

A large number of parasites infect fish, but only a few cause illnesses in humans. Many marine and freshwater fishes serve as a source of medically important parasitic zoonoses that include trematodiasis, cestodiasis, and nematodiasis. Some of these infections are highly pathogenic. These diseases are mainly acquired through eating raw or under cooked fish. Generally, fish can either be intermediate host of parasites involving man as the definitive host, or harbor larval parasites of other animals which can invade human tissues. However, the larval stages of a few species of parasite can mature both in animals and man. The reported incidence of these fish zoonoses has increased in recent years due to the development of improved diagnosis, increase in raw fish consumption in those countries in which such dishes have commonly been eaten, increased consumption elsewhere of regional fish dishes based on raw or poorly processed fish, the growth in the international market in fish and fish products, and the remarkable development of aquaculture.

Trematodiasis

Trematodiasis is the infection of humans by trematode parasites. Among the fish parasitic diseases, infections by digenetic trematodes are the most common. A considerable number of digenean metacercaria from fish may infect humans. The disease is important in Southeast Asia and the Far East where many people are dependent on freshwater fish as the major source of protein. The most significant of these digeneans are perhaps *Clonorchis sinensis, Opisthorchis viverrini*, and *Opisthorchis felineus*. A large number of freshwater fish species can transmit the infective trematode metacercariae with fish belonging to the Cyprinidae (carps) being the most common. Farmed fish of a variety of species have also been shown to be hosts of trematode parasites.

Clonorchiasis caused by the liver flukes *C. sinensis* is endemic in South China, Taiwan, South Korea, and North Vietnam. It is estimated that 35 million persons globally could be infected by *C. sinensis*, including 15 million in China. The disease is being associated with biliary obstruction leading to hepatic necrosis, cirrhosis, and portal hypertension, in heavy infections. The parasite may also locate in pancreatic ducts, causing acute obstructive pancreatitis, a most painful condition. Opisthorchiasis caused by *O. viverrini* is endemic in Thailand, Lao, Cambodia, and Central

Vietnam. It is estimated 6 million humans in Asia may be infected with *O. viverrini*. Human infection due to *O. felineus* is found in Russia and countries of Central Europe.

Cestodiasis

There are relatively few cases of fish-borne cestode infections in man. The cestodes that mature in the small intestine of man are not pathogenic, and diseases are never fatal. Diphyllobothriasis is the major cestodiasis transmitted by freshwater, marine, and anadromous fishes. The disease is caused by pseudophyllid cestodes belonging to the genus *Diphyllobothrium*. At least 13 species of the cestode genus *Diphyllobothrium* have been recognised from humans. The genus is found in fish, mammal, and avian hosts, and is usually associated with cold-water habitats. The species most often reported from humans is *D. latum*, which is relatively common in the Baltic region, the European Alps, eastern Russia and Japan. It is considered a mild disease; persons infected with the tapeworm may often be symptomless, in others it may cause diarrhea, abdominal pain, and anemia. Recent estimates indicate that approximately 20 million individuals could be affected by the disease.

Nematodiasis

Fish-borne nematodiasis are generally caused by the incidental infestation of man with nematodes whose natural definite hosts are marine mammals, birds, pigs, or other animals. Freshwater, brackish, or marine fishes are the second intermediate host. In most infections, the worms can only survive for a limited period after the initial invasion of the gastrointestinal tract. The method of infection is by ingesting the infective larvae which are located in the muscles, intestine, or viscera of fish.

Capillariasis is caused by nematode *Capillaria philippinensis*. The disease was originally presumed to be an indigeneous disease of the Philippines, where an epidemic was first recorded in 1967. Subsequently, the disease was also found in Thailand, Japan, Taiwan, Indonesia, Korea, Iran, Egypt, and India. Freshwater fish may be important as a source of infection of humans with this nematode. The adult worms are found in the gut of humans, where they can cause a severe and even fatal illness.

Gnathostomiasis is caused by members of the genus *Gnathostoma* who undergo visceral larval migration. The disease occurs in Southeast Asia, China, Japan, Korea, the Indian subcontinent, and Middle East. Its life cycle is complex involving intermediate (crustaceans and fishes), paratenic (piscivorous birds, reptilian, and small mammals), and final hosts (wild and domestic animals). Man is considered an accidental host in whom the parasite can cause a wide clinical picture, internal or external, where the condition 'larva migrans' is one of the known symptoms.

Anisakiasis refers to infection by larval ascaridoid nematodes if ingested with raw or lightly cured fish. The genera involved are *Anisakis*, *Pseudoterranova* and *Contracaecum*. Their normal definitive hosts are marine mammals. Larvae (located in squids and marine fish) can invade the gastrointestinal tract of man, causing an eosinophilic granuloma syndrome. In Europe, it has also been referred to as the 'herring worm' disease. These nematodes cannot mature in humans, but may cause a severe allergic reaction with granulomatosis of the stomach wall.

A. simplex causes an acute or chronic infection that may lead to abdominal pain, nausea, vomiting, and diarrhea. Some patients develop syndromes exhibiting clinical manifestations of allergy following infection or following consumption of dead larvae. The incidence of the disease varies widely among countries, with Japan reported as having the highest incidence.

Permissions

All chapters in this book are published with permission under the Creative Commons Attribution Share Alike License or equivalent. Every chapter published in this book has been scrutinized by our experts. Their significance has been extensively debated. The topics covered herein carry significant information for a comprehensive understanding. They may even be implemented as practical applications or may be referred to as a beginning point for further studies.

We would like to thank the editorial team for lending their expertise to make the book truly unique. They have played a crucial role in the development of this book. Without their invaluable contributions this book wouldn't have been possible. They have made vital efforts to compile up to date information on the varied aspects of this subject to make this book a valuable addition to the collection of many professionals and students.

This book was conceptualized with the vision of imparting up-to-date and integrated information in this field. To ensure the same, a matchless editorial board was set up. Every individual on the board went through rigorous rounds of assessment to prove their worth. After which they invested a large part of their time researching and compiling the most relevant data for our readers.

The editorial board has been involved in producing this book since its inception. They have spent rigorous hours researching and exploring the diverse topics which have resulted in the successful publishing of this book. They have passed on their knowledge of decades through this book. To expedite this challenging task, the publisher supported the team at every step. A small team of assistant editors was also appointed to further simplify the editing procedure and attain best results for the readers.

Apart from the editorial board, the designing team has also invested a significant amount of their time in understanding the subject and creating the most relevant covers. They scrutinized every image to scout for the most suitable representation of the subject and create an appropriate cover for the book.

The publishing team has been an ardent support to the editorial, designing and production team. Their endless efforts to recruit the best for this project, has resulted in the accomplishment of this book. They are a veteran in the field of academics and their pool of knowledge is as vast as their experience in printing. Their expertise and guidance has proved useful at every step. Their uncompromising quality standards have made this book an exceptional effort. Their encouragement from time to time has been an inspiration for everyone.

The publisher and the editorial board hope that this book will prove to be a valuable piece of knowledge for students, practitioners and scholars across the globe.

Index

A

Actinopterygians, 43-44, 56, 58

Actinopterygii, 43-44, 56, 103

Anal Fin, 9, 55, 98, 102, 106-108, 116, 163, 173, 183

Anchovies, 37, 44, 96, 197

Atlantic Cod, 37, 138, 141-142, 153

B

Basal Plate, 90

Bathypelagic, 4, 151

Benthic, 4, 177-179

Blastomere, 23

Bluefin Tuna, 7, 68

C

Cartilaginous Structures, 45

Catfishes, 6, 8-9, 11-12, 16, 44, 96, 147, 160

Caudal Fin, 10, 12, 39, 57, 93, 97, 100-102, 104-106, 109, 114, 117-119

Caudal Fins, 39, 103, 114, 119

Caudal Peduncle, 81, 95-96, 101-102

Chimaeras, 13, 45-46, 50-54, 63-64, 87, 90, 96, 104, 147, 160, 162

Chondrichthyes, 2, 43, 45-46, 54-55, 67, 104

Chordata, 1

Chromatophores, 11

Cichlids, 9, 43-44, 60, 65, 74, 82, 88, 156, 164, 174, 179, 181, 183

Cosminelike Layer, 11

Cosmoid Scales, 11, 93-95

Ctenoid Scale, 92

Cuttlefish, 190

Cycloid Scale, 91

Cyclostomata, 42

D

Dorsal Fins, 10, 41, 55-56, 97, 103, 108, 115-116, 118, 137

E

Ecological Niche, 6, 82

Efferent, 14, 62

Embryology, 1, 110

Embryonic Shield, 26-29

Endochondral Bone, 44-45

Endoskeleton, 190

Erythrophores, 11

Exoskeleton, 104

F

Fish Locomotion, 6, 113-114, 117, 120

Fish Populations, 1, 197, 199

Fisheries Management, 2

Fishery, 47, 199

G

Gene Expression, 3, 79, 138

Gill Arches, 10, 13-14, 50, 60-62, 80, 83, 86-87, 110

Gill Cover, 13

Gill Filaments, 13, 52, 179, 212-213, 215-216

Gill Slits, 1-2, 52, 61-63, 85-88, 148

Gnathostomata, 41

Guitarfishes, 46, 52, 55

H

Hagfish, 38-40, 43, 63, 65, 71, 88, 126, 172

Hairyfish, 10

I

Ichthyology, 2

Iridocytes, 11

Isopedine, 11, 94

J

Jawless Fish, 43, 61, 71, 75, 83-84, 86, 126, 162

L

Lampreys, 1-2, 5, 10, 18, 38-43, 63-64, 76, 83, 85, 88, 110, 136, 147, 172

Larval Food, 5

Lobe-finned Fishes, 55-56, 94, 103-104

Lungfishes, 13, 51, 55-58, 64, 77, 94, 102, 111, 147

M

Melanophores, 11

Milt, 8, 108, 163, 170-171, 173, 186

Monophyletic Group, 44, 56

Moray Eels, 79, 121, 157
Myxinidae, 41
Myxiniformes, 38, 41

N
Neopterygii, 44
Notochord, 1, 10, 27-29

O
Olfaction, 16, 32, 72, 128, 167
Osmoregulation, 66-67, 148, 204
Osmotic Pressure, 15
Osteichthyes, 43-44, 54, 56, 58, 102
Otoliths, 44, 72, 127
Ovoviviparous Sharks, 53
Oxygenated Blood, 14, 124

P
Paddlefish, 130
Paleontology, 4, 10, 160
Pectoral Fins, 7-8, 10, 52, 97-98, 104-105, 107-109, 111, 114-117, 119-120, 122, 124, 141, 215
Pelagic, 4-6, 8, 33, 57-58, 62, 116, 177-178, 180, 184, 197
Pelvic Fin, 51, 109, 111
Petromyzon Marinus, 40, 136
Petromyzontiformes, 41
Phylum Chordata, 1
Physostomous, 13, 123
Pickerel, 44
Placoid, 2, 11, 54, 89-90, 94, 96
Placoid Scales, 11, 54, 89-90, 94, 96
Poikilothermic Animals, 1
Posterior Termination, 9
Protopterus, 13, 57

R
Radial Bones, 10
Ray-finned Fish, 43-44, 102-103, 124

S
Salmo Salar, 138
Sarcopterygians, 43-44, 56, 58, 94
Sawfishes, 46, 50, 96
Sea Bass, 47, 136
Semelparous, 172-173, 190
Sensory Receptors, 10, 132, 135, 149
Shellfish, 1
Sportfishing, 3, 43
Streamlined Body, 6, 69, 120
Superclass Agnatha, 38, 41-43
Swim Bladder, 2, 13, 44, 54, 60, 70-72, 88, 118, 121-128, 220

T
Teleostei, 44
Teleosts, 8, 10-11, 13, 15, 45-46, 62, 64, 68, 73, 83, 86, 90, 94, 162, 164, 172
Thresher Sharks, 50, 68, 97 Trout, 2, 7, 10, 17-18, 30-31, 37-38, 44, 73, 100, 114, 131, 137-145, 153, 157, 174-175, 179, 222, 224-226, 229-230

U
Urogenital Opening, 8

V
Vital Organs, 9, 126
Vitrodentine, 11, 90, 93, 95
Viviparous, 8, 53, 74-75, 165, 173, 184, 201-203, 228

W
Weberian Ossicles, 123

CPSIA information can be obtained
at www.ICGtesting.com
Printed in the USA
LVHW062014100122
708212LV00002B/234

9 781639 875528